The New RIGHT STUFF

Using Space to Bring Out the Best in You

Loretta Hidalgo Whitesides

Revision 21, May 2021

For Richard Condon

Thank you for all the love, coaching, support and wisdom you gave so generously
You helped me fly.

And you helped me craft a course so I could pass on all that was given to me
Thank you for helping me Leap!

Contents

Introduction 9

1. Start Where You Are 1

Welcome to Your New Right Stuff Training 1

The Most Complex Subsystem of the Spaceship is You 2

What is The New Right Stuff? 3

A World That Works 4

What Is Your Mission? 6

Dream Big 7

Ask for Help 8

Never Give Up 10

Who's With You? 11

Optimists Have More Fun 12

The Pain of Regret 14

Seeing Your Blind Spots 14

What If I Spent All My Energy Being Awesome? 16

The Power of Star Wars 17

The Hero's Journey 19

2. Learning Your Power 23

The New Right Stuff 25

You are Literally One in a Million 27

Be Like a Cat 28

You Can't Observe a System Without Affecting It 28

What Is Your Challenge? 30

Listen for the Gold 30

It's Not About The Nail 31

Speaking with Respect 32

You're a Sophisticated Tone and Mood Sensor 33

Speaking With Love 35

Be a Jedi Audience or Meeting Attendee.....36
Never Say Never.....37
Jedi Patience.....38
A Jedi Craves Not These Things.....39
Building the Culture to Go Along With the Spaceships.....40
It Takes A Village.....42
Connecting With The Force.....43
Keeping The Big Picture Present.....45
Read a Book.....46
Finding a Coach.....47
Ride the Horse the Direction It is Going.....48
Daily Action.....49
Review.....51
3. The Dark Side.....53
Facing Our Dark Side.....54
Going Inside the Cave.....57
The Way of the Jedi Warrior.....58
Don't Throw Anyone Under The Bus.....59
Transcend 'Us Versus Them' Thinking.....61
The Screen Saver of Your Mind.....62
Don't Type Things You Don't Want Forwarded.....63
Giving Up Sarcasm.....63
Stress.....64
Dissatisfaction.....65
Same Boss, Different Company.....67
Analysis Paralysis.....68
Anger.....69
Fear.....71
Aggression.....72
Arrogance.....73

Failure 74
Criticism 75
Jealousy 77
Coughing Up Furballs 79
Learning What Triggers Furballs 80
Being Responsible for Your Shortfalls 82
Taboos 83
Suicide and Mental Health 84
Homosexuality and Sexuality 86
Making Peace With Your Pieces 87
4. Do or Do Not, There Is No Try 91
The Four Agreements 92
Do or Do Not, There Is No Try 94
When My Integrity is Out, Drama Follows 96
Leave It Cleaner Than You Found It 98
Power of Humor and Play 99
Using Inclusive Language 100
Privilege is Blind 102
It All Comes Down to Relationships 103
The Org Chart and the REAL Org Chart 104
Answer the Question That is Asked 104
Being on Time 106
"Yes, and…" Speaking 107
Remembering People's Names 108
Remembering More Than Their Name 110
Help People Remember Your Name 110
Help Others Remember Names 111
Graciously Accept Compliments 112
Generously Acknowledge People 113
Ask To Be Acknowledged 114

A New Hope....115

5. Courageous Conversations....119

Writing Your Obituary....120

Restoring a Relationship....121

Were My Insecurities Triggered?....121

When to Speak Up....122

The Power of Being Kind....124

The Power of Taking Responsibility....125

The Power of Forgiveness....126

The Power of 'I' Statements....127

Conversation Pitfall: Triggering Defensiveness....128

Where Do I Start?....130

Step 1: Beginning the Conversation....131

Step 2: Get Interested In Them....131

Step 3: Acknowledge Their View....133

Step 4: Share Your Perspective....135

Step 5: Brainstorm Solutions....136

Step 6: Practice, Practice, Practice....137

What If They Really ARE Wrong?....138

6. Jedi Ways of Being....141

Living in World 2....141

Finding Your Niche....142

Running a Lean Meeting....144

Happy People Do Better Work....146

Be Willing to Say "I Don't Know"....147

Receptivity: Tuning in to the Universe....148

Thinking Original Thoughts....150

Appreciating Where You Get to Be....150

Live Like You Are On A Spaceship (You Are)....151

Be Yourself, No Matter What They Say....153

Being Charismatic....153
Being Audacious....154
Being Fun....158
Being Creative....159
Your Attitude Determines Your Altitude....160

7. What You Came To Earth to Do....165

Redeem Your Parents....166
It's Time....168
What Is Your Mission?....169
Finding Your Purpose....172
Do the Most Important Things First....174
Ask for What You Really Want....175
"So, What Do You Do?"....176
Get Bigger Problems....177
Imagination is More Important Than Knowledge....178
The Future Doesn't Just Happen, It Has to Be Created....179
The Real Reasons We Go Into Space....179
Going to Space is Like Having a Baby....181
We Need to Learn to Be Better Parents....181
Vulnerability Is The New "Right Stuff"....183
Building Your Own Jedi Circle....184
Using the Force of Knowledge and Defense, Never for Attack....185
Passing It On....186
Follow Your Dreams and You Will Give Others the Courage to Follow Theirs....187

Acknowledgements....192
Resources....194
Index....195

Introduction

This book is designed to be workbook, a guide, a course for you on your Hero's Journey.

Take the time to work each chapter's exercises and homework before moving on to the next chapter. I recommend a week for each if you can. You can work it with a buddy, a book club, with our live or recorded SpaceKind Trainings or on your own.

Whatever you do, I wish you luck on your quest. You are setting off to write your own myth. It can lead to unexpected, challenging and ultimately liberating places.

In the end you will return home a little closer to who you have always wanted to be.

Thank you for having the courage to start.

Loretta Hidalgo Whitesides

JON MARRO

• CHAPTER 1 •

Start Where You Are

Welcome to Your New Right Stuff Training

You may feel called to greatness, called to make a profound difference in the world, called to be of service to humanity.

Or you may not be sure exactly what your calling is, or you may be daunted at the idea of undertaking it. Wherever you are, you are in the right place.

We are glad you are here. Welcome home.

It is not a mistake you are reading this book. You know it is time to answer the call. To do that, you are going to have to say YES to this journey. You don't have to of course. You can save it for another day. You can return to the Ordinary World. You can play it safe.

But if you are up for a dive into the Special World, if you are up for challenge, growth and ecstasy, all you have to do is…

…turn the page.

The Most Complex Subsystem of the Spaceship is You

In human spaceflight, the things that pass for normal in a typical work environment kill people.

One of the hardest parts for me of the Space Shuttle Columbia accident in 2003 was that it meant we hadn't learned the lessons deeply enough from the Space Shuttle Challenger accident 17 years before—lessons I felt it was my responsibility to teach.

Being in Mojave in October 2014 watching the fourth powered test flight of Virgin Galactic's SpaceShipTwo was even harder. That flight ended when the ship broke up just after the transition to supersonic flight killing one of the two test pilots onboard.

Spaceflight accidents are not just engineering mistakes. They often also involve painfully human shortcomings.

We must never forget the human failings that led to those accidents. Hubris (excessive pride or confidence), unwillingness to speak up, unwillingness to listen, unwillingness to ask for help, unwillingness to accept help or even admit we need help, ego and even basic personality conflicts. Any of those can kill people, crash spacecraft, and sideline our dreams.

These very human shortcomings are regrettably quite common in modern office environments and so no one really questions them. Comics and TV shows like *Dilbert* and *The Office* parody them, but the dysfunction of humans in work environments is largely left unchecked.

We have to hold ourselves to a much higher standard if we want to survive in the harshness of space. What passes for normal in a modern office needs to become completely unacceptable.

The space industry has tried to put in more and more layers of safety, requirements and process to compensate for the lack of trust, communication, excellence, humility, and teamwork in a typical office environment. But no matter how many processes you impose, if the individuals and the team are not operating from integrity, love, trust and communication, giving it all for the higher good, and being willing to be uncomfortable—not just with long hours, but with sharing their concerns or mistakes, or working through their differences with a coworker, our teams will still not be working at their highest capacity. Moreover, we will not achieve what we are capable of as a species.

This is where your New Right Stuff training comes in. This book is designed to help YOU to be a force for good in your team, your company, your industry and your species. When enough of us really take this on, there will be a noticeable shift in the future of humanity.

What is The New Right Stuff?

In the 1960's you had "The Right Stuff" if you were smart, willing to be physically at risk and could stay cool under pressure.

Now we need people willing to put themselves at risk, not just physically, but emotionally. We need people who are smart and able to connect with others. We need people who can stay calm under pressure and also be honorable, vulnerable and inspiring.

I call this the New Right Stuff.

Do you know anyone like this? Those rare and magical people who can:

-Diffuse tense situations

-Work with anyone

-Communicate in a way that honors everyone

-Share useful insights

-Inspire people to be their highest selves

-Produce extraordinary results

Do you see these traits in anyone you know? Are you interested in developing them?

The truth is that our whole lives would be amazing if we were being this way all the time. This book is designed to help you have an awesome life, to fulfill what you came to Earth to do and to be someone with the New Right Stuff.

A World That Works

This is also bigger than just getting your life to work. This is about getting *all* of life to work. Imagine a world where all people are inspired, responsible and engaged in fulfilling their life mission.

Water would be clean, kids would be safe and fed, people would work through their differences peacefully and respectfully.

We are a long way from that world, yet that is the world we must be committed to creating. Why? Because things would get more interesting when we are closer to fulfilling on our potential as a species. What could we create? What would become possible? That is the future we must build together.

As a kid I saw many images of what the future would look like—gleaming cities, high tech gadgets, flying cars. In all of these images the faces of the people were always beaming. It's as if they are saying, "Wow! The future is going to be so great!"

Now I live in a future that in many ways is beyond what most of us imagined in the 1970's. In 2019, a girl in Bangladesh can access more of the world's information from a device in thc palm of her hand on a moments notice than the President of the United States could in the early 1980's. I can video call friends in nearly any country on the planet for free and post ideas that anyone with an internet connection could read, even in a different language.

But, contrary to our hopes, great leaps forward in technology, life expectancy, literacy, medicine, wealth and leisure time have not lead to increases in our sense of happiness, well being, or fulfillment. Suicide, depression and divorce rates are still too high. Anger, fear and aggression are still too common. There is more to building an amazing future than improving our physical world.

We must also work on *ourselve*s as we improve our technology. Think through what you hope new technology or new advances in science will unlock for our experience of life and then think through what we will need beyond the tech to “make it so.”

It’s time for us to grow up as a species. It’s time for us to be a match for our technological progress.

Since change starts with us, we’ll start by looking at what your mission is. What role can you play in all of this?

What Is Your Mission?

The first part of this journey is you discovering and articulating your mission. **What did you come to Earth to do and why?** It is as unique to you as your fingerprint and will give you the power and perspective to carry you through many frustrations and challenges.

It is also ok if you don’t know what it is or have never felt particularly called to a specific mission. That just means you get to create it now, based on what inspires you in this moment.

“Using Space to Bring Out the Best in Humanity” is my mission. It is what I came to Earth to do. It is what I want my contribution to the world to be. Part of that involves helping people to find their calling — their life purpose. What is yours? What is your mission? What did you come to Earth to do? Do you remember?

What mission comes to you as you read this? Jot down your first impression. Write down any other possibilities that come to mind. You can always refine it or even change it later.

____

____

____

____

____

Dream Big

Often the college students I work with will have dreams like, "to build rockets" or to "work at NASA" or even, "to be an astronaut!" I tell them that that dream is *way* too small—they will be able to knock that one out in just a few years, and then what? What do you want the rest of your life to be about? Even if you dream of being an astronaut, *why* do you want to be an astronaut? How will you use that experience to make a difference for others?

When I speak at universities, I always encourage the students to "Dream Big." I am pretty straight with them. Sometimes being in the workforce can be challenging. You may have long meetings to sit through, frustrating colleagues, and technical setbacks. Unless you have something much bigger than yourself to look to, it will be hard to keep perspective and continue to seek out ways to make your work worthwhile.

You want to have a dream big enough that it will take your whole life to fulfill it. If your dream is something that a professor would tell you is impossible, then you are on the right track! The role of the young is to make real that which previous generations thought impossible: heavier-than-air flight, heart transplants, running a four-minute mile, walking on the moon.

My dream is to use space to alter the trajectory of humankind. I want us to be more Jedi-like, more peaceful, like the society we see in Star Trek. Some people want to help build the spaceships we see in science fiction; I want to build the society we see there. I am willing to devote my whole life to that.

What is your impossible dream?

____

____

____

____

____

Ask for Help

The first thing you notice when you take on the impossible is that it isn't exactly easy! You will most likely not be able to accomplish it alone. When the United States put a man on the moon in 1969 it took nearly 400,000 people to make that happen. If you want to put a human on Mars, take a picture of the nearest Earth-like planet, or travel faster than the speed of light, you are going to have to ask for help. Whether you are asking for funding, or collaboration, or someone to join your project, it all involves *asking*.

Now, I know that I am not good at asking for help. Many of us got to where we are by being confident, independent and self-motivated. You can do well in life just on your own merit, so often the skills to ask for help are underdeveloped. We would rather take a Lyft or a train to the airport than ask someone to give us a ride. We will stay up late collating papers for a presentation the next day rather than ask someone else for help.

But now is a good time to start practicing asking for help. Ask someone to volunteer on your project, ask someone to write you a LinkedIn recommendation, ask someone for a donation for a cause you support. Because if you don't ask, the answer is always going to be no.

At some point you may have to stand before Congress or potential investors and ask for a billion dollars for your idea; you might as well start practicing asking for big things now.

What are two things you could ask for?

____

____

____

____

Never Give Up

As you start to work on the impossible, you will have setbacks, challenges, rejections and days you just "don't want to." Your job is to make sure that when you get frustrated, upset or disappointed that you do not give up. EVERY time you are upset, frustrated, or disappointed, remember that not being stopped by that is part of your training to be extraordinary.

Nearly every extraordinary accomplishment has an equally harrowing story of rejection and failure to go along with it; from Lincoln's nine lost elections, to Edison's 10,000 light bulbs that exploded, to The Wright Brothers being told that "heavier than air flight is not possible" by Lord Kelvin just a few years before Kitty Hawk.

To be someone who can make the impossible happen, you must become a Master at dealing with upset, frustration, disappointment, rejection, failure and hopelessness.

It takes about 10,000 hours to become a Master at anything, be it tennis, piano or quantum physics. And just as the masters of these skills must practice their craft consistently week after week, we too must continually push ourselves to face upsets, frustrations, and resignation and not let it stop us.

So the next time you are angry with a colleague for treating you poorly, or upset about making a mistake, or disappointed the demonstration didn't work, just take a deep breath and know that this is the perfect training for your Mars mission, your CEO job, or running Luna City. This is the preparation you need to better fulfill your mission.

Each time you go there and let go of the upset, the frustration, the disappointment, and come back to your dream, you are building the muscle of resilience, of knowing yourself as someone who can get up and go to work in the face of anything.

Those are the people who make a dent in the universe.

Time to start training.

Who's With You?

Never giving up doesn't require you to be resilient all by yourself. The reason I don't give up is because I have a whole village backing me up. I even have a new mantra, "When I can, I do; When I can't, I call a friend."

When I am upset, frustrated, or numb, I immediately call for backup. My own negative thoughts can spiral downward pretty quickly. So I remind myself, "Your head can be a bad neighborhood; don't go in there alone!" One of your first jobs is to build a great circle of people you can lean on when you need support.

Start with others who share your mission, or who understand you on a really profound level, or even others who are reading this book. You want people who will only listen to you as awesome and capable and will not let you marinate in self-pity or self-doubt.

Who else do you want with you on this journey? Make a list of at least two people you want to support you in being your best self. Ask one to be your buddy and read along with you. Share with them the thoughts and ideas you have as you read the book.

Just like having a gym buddy makes you more likely to go to the gym, having a New Right Stuff buddy will help you become the person you have always wanted to be.

Optimists Have More Fun

Pessimists are optimists who have gotten their heart broken. They would rather be right that things are going to go wrong, than be disappointed again.

An optimist says, "I am going to get into Harvard!" Then when they don't they say, "I guess the universe has something better planned for me!"

A pessimist says, "I'll never get into Harvard." Then when they are not admitted they say, "See, I told you so."

Being right is small solace. And being pessimistic can also negatively affect the outcome. If you go into an interview (or a date!) with the thought that "they will never pick me," that message gets through; it is in the unsaid.

People tell me, "But I didn't *tell* them I didn't think I was a good fit for the job." You don't have to—our words are only a fraction of what we are communicating. Our body language, who we are being, and the tone and inflection of our voice often scream louder than our words. That is what the people around us hear.

Being an optimist can actually improve the outcome. If you learn the Jedi mind trick to avoid being crushed by disappointment, you will be freed up to be optimistic about anything.

So how do you handle disappointment? You look for the gift in its hands.

Every disappointment is part of your training. What does the universe want you to learn from this? How can you use it to make you better? What does this disappointment make room for? "Well, now I will have plenty of time to study Russian!"

You can also say that it happened for a reason. There is something better coming along. You need to be at Caltech to meet that mentor who will open up important doors for your dreams. Trust that the twists and turns of your life are all perfectly timed to land you where you need to be.

If you embrace everything that is happening as the perfect thing, you no longer need to fear or avoid being disappointed. You are free to be as optimistic as you want—and to have more fun.

The Pain of Regret

I often resist doing the hardest work because I know it will hurt. Whether it's running during the dead of winter, having to raise a painful subject with a friend, or giving up my down time after work to move my side projects forward.

Anything that matters, anything worth having is going to take something from you to make real. As Jim Rohn says:

"There are two types of pain you will go through in life, the pain of discipline and the pain of regret.

Discipline weighs ounces while regret weighs tons."

This book is about getting you to do the hard work of discipline so you can avoid the pain of regret.

It's your choice.

Seeing Your Blind Spots

Clasp your hands together with your fingers intertwined.

Notice how you naturally have a way that you are most comfortable doing this.

Now, unclasp your hands and re-clasp them with the opposite index finger in front. How does that feel to you? For most people it feels pretty awkward and strange.

You have many things like this all over your life. Things you do automatically, unconsciously, without even realizing it is *your* way of doing things (and that there are equally valid alternate ways to do it).

These habits impact your life way more than how you intertwine your fingers. Habits like getting defensive when you feel criticized, being sarcastic with people you think are not as smart as you, or avoiding your boss so you can avoid confrontation.

Usually others can see our blind spots much more clearly than we can. That is why having people you can sit down with and ask them to level with you, hearing them say what your strengths and weaknesses are, is so valuable (we will be doing that later in the course). They can tell us things it could take us years to uncover on our own (and usually only after painful trial and error).

Make sure to make the safest possible space for them to be honest and for you to listen without defensiveness or comment. You really need to resist the urge to say, “No, I don’t do that,” and then disregard the input.

These things are our blind spots, our ways of operating in the world. They will have a bigger impact on whether you will be promoted than where you went to school and sometimes even how smart you are. Be open to exploring them more as we go on…

What If I Spent All My Energy Being Awesome?

During an intense coaching weekend I saw a big blind spot I have. I spend about 80% of my energy on being angry, being critical of others or being hard on myself. That is a staggering amount.

While those things are easy, deeply engrained and give me a nice sense of superiority, they totally and completely get in the way of my goals and what matters most to me.

How can I bring out the best in humanity when I am screaming about the back door being left open in the dead of winter? How can I promote a world where we all work together when I am judging someone for being too dense to figure out what 200 divided by 8 is? And most crushingly how can I contribute my light to the world when I am down on myself anytime I feel low?

I may be "busy" and I may have a lot of my plate, but if I could reliably give up being angry I would have a lot of time back. If I could give up being critical and condescending, I would have a lot more energy. And most critically, if I could give up being hard on myself, I would be happier, more effective and more present.

Success isn't about cramming more things into your life and sleeping less. It's about taking away habits, patterns, and dead weight that don't serve us. It is being lighter that frees us up for greatness.

The Power of Star Wars

In this book I have a lot of *Star Wars* References. *Star Wars* is more than a movie for me. I hold *Star Wars* with a deep reverence.

The Force, Yoda, Obi Wan, powerful and brash Princess Leia, the twin setting suns of Tattoine, the incredible soundtrack, the cynical Han Solo, the starry-eyed Luke Skywalker, the quotable lines, the on-screen chemistry, the willingness to risk everything for the cause.

Most sacred of all of course is the on-screen portrayal of what it means to be a Jedi.

Obi Wan demonstrates what it looks like to live peacefully, honorably, without anger or malice. He moves gracefully, decisively, purposefully, and is not stopped by fear. He is not afraid to expect a lot from people or to show his faith in the Force.

Yoda is the most compelling character I have ever seen on screen. He can use humor to make a point, as he does upon meeting Luke Skywalker, but he is even more powerful when he channels pure Jedi, closes his eyes, draws a deep breath and lifts an X-Wing fighter out of a swamp. He speaks slowly, each word measured out for maximum impact. He is very present. He teaches us the importance of giving up anger, fear, and aggression as much as he teaches Luke. He is our teacher, for most people the wisest influence they may ever encounter.

The Jedi are one of the most powerful secular moral teachers we have on this planet. Putting their teachings in front of billions of people, could train a generation to be peaceful, patient, loving, and centered.

How?

By modeling what it looks like. By giving the world tools and dialogue that they can quote over and over to remind them to let go of anger, fear and aggression.

You honor what it means to be a Jedi by upholding it as something sacred, as something worth honoring.

"Human passions," says the historian Durkheim, "stop only before a moral power they respect."

The Force, and the way of the Jedi, is one of the few things that are still sacred in large parts of our culture. Jedi are compelling. They have a lot of on-screen charisma and presence. They are always in the moment.

They are calm. They make you think. They make you want to be a better person. They call for your best and highest self. That is the role of a Jedi.

For me, they are the spiritual leaders of even the most secular parts of the planet. They are the opportunity to train billions in powerful memes about letting go, about drawing on forces greater than yourself, about doing the work to train yourself to be of service to humanity, about never giving up on anyone, about anything being possible, about light being more powerful than dark, about finding a teacher, about the value of patience and the value of being present, the importance of facing your fears, and the joy of being part of a noble cause and a noble team.

The Jedi model equanimity (calm) on screen, show what it looks like not to get ruffled—to laugh at themselves as Yoda does, to have a deep reverence for the Force, to be powerful and humble at the same time, to not take things personally and to act from what is best to restore order to the Galaxy.

There is a lot we can learn from them.

The Hero's Journey

Star War's creator George Lucas was heavily influenced by Joseph Campbell's book, *The Hero With a Thousand Faces*. The book is about the "Hero's Journey," the transformative journey that is common to hero myths and stories from nearly every culture from the Greeks and Romans to today.

The journey we will take in this book is also designed to follow the path of the hero. From answering the call to adventure, to meeting your mentor, to crossing the threshold to the Special World, to being challenged, to training and preparation, to an ordeal where the hero must face death, to revelation, to a victory against the odds and then the journey back home to use what you have learned in the service of others. That is what heroes throughout the ages have done and this is what we will do.

Homework:

1). Start a journal or write in the margins of this book. Write down the first draft of what your Mission is. Include what you want to accomplish by the end of this training.

2) Answer the Call. Notice any opportunities that present themselves to work on what you said you wanted to work on.
Be sure to say YES!

3). Ask someone for help. You can ask for help on a work project, a backyard project, a volunteer project or anything in between.

4). Find a buddy you can take this journey with. Talk about each other's Missions, this chapter, and what you are working on/noticing in your lives.

JON MARRO

• CHAPTER 2 •

Learning Your Power

The first step in learning your power is to articulate what it is that you are most passionate about. What do you want to see happen in your lifetime? What role do you want to play in bringing that future to reality?

I am pretty clear that we all came here with a mission. My job is to help you identify what your role is and to fulfill it to the best of your ability. If all of us were freed up to be 100% who we are meant to be, fulfilling on what we came here to do, and wailing— playing full out—what an amazing world this would be!

The people called to work on climate change would work on stabilizing CO2 levels in Earth's atmosphere. The people called to work on supersonic transport would figure out how to make it feasible and sustainable to get around the world faster. The people called to work on empowering kids to go into science and engineering would be doing that. It would all get done because the people who are here to work on that would be playing full out and getting it done.

It starts with writing it down (as you did in the previous section) and being honest with yourself about what your mission is.

The next step is to share it with people. Put it in your email signature, put it on your Facebook profile, print it out (or paint it...) and put it on your desk. Let the world know who you are and what you are here to do. Who knows; someone else may be looking to connect with the very thing you are here to do.

We will then work on eliminating anything that is in the way of you stepping up and being that. We will look at what might be slowing you down at work, bogging you down in life, or is in any way interfering with what you came to Earth to do.

Being vocal and being unabashedly who you are inspires others to put their ideas out there and be who they are too.

Follow your dreams and you will give others the courage to follow theirs.

The New Right Stuff

It might seem scary to share your mission with the world. It does to me. But to really alter what's possible for human beings, to really make the future something beyond our wildest dreams not just in terms of technology, but also in terms of our happiness, our comfort in our own skins, our fulfillment, our level of intimacy with the people we love most, it is going to take risk.

Just as having the Right Stuff meant being willing to put their lives at risk, the New Right Stuff is about sharing your inner most life. For many that is much more terrifying than being physically at risk.

Making a profound difference for the world and being role models of what is possible for human beings will mean our having to be vulnerable. It will take saying, "I don't know how, can you help me?" and "I am sorry, I shouldn't have spoken to you that way," and, "That was terrifying, I nearly lost everything that ever mattered to me." It will take us supporting one another and being real with one another.

The downside of trying to live up to the old "Right Stuff " is that we end up feeling like we have to pretend that we "have it all together." Then we go home to hide in our safe space and try to numb out the disappointment, sadness, or fear we are carrying. We all numb ourselves in different ways. Some of us numb ourselves by working out, others by shopping, watching screens or something else. That culture has led to a lot of alcohol abuse, marital infidelity, and stress.

Still, we have a lot to learn from the incredible accomplishments of the Apollo era. The 400,000 people who worked together on the Apollo Program had to be awesome at what they did to make it all work 238,900 miles from home. I think Apollo was absolutely amazing and there were a number of extraordinary leaders from that time like Gene Kranz, Chris Kraft and Wernher von Braun.

What I am asking is, can we build on what they did? Can we do it even better? Can we do it without breaking up marriages and driving people to drink?

What if we could stand on the shoulders of giants, build on their extraordinary, inspiring achievements and add in taking care of people, open communication, respect and smoke-free conference rooms? Imagine the kind of program that would be. Actually I suppose we don't have to. Gene Roddenberry, the creator of Star Trek, did it for us.

Now, just the way cell phone and 3D printers are turning Star Trek technological visions into fact, it is time for us to help make Gene's vision for the culture of the future a reality, too.

It starts with You.

You are Literally One in a Million

Many space scientists, engineers, and advocates feel as if they are overstepping their bounds to really swing out and boldly lead the way or take a stand, or start an initiative, a project, or a company.

But who are you not to, really? The world is in dire need of leadership right now. They are desperate for solutions, for guidance, for answers. Who are you to hold back what you have to offer?

Just to give you a sense of how unique you are: there are 7 billion people living on this planet, 330 million people live in the United States, 67 million are college graduates. Five million of them work in science and engineering. Some 18,000 people work at NASA. Less than 7,000 people work in your area of expertise with your experiences and commitments. You are literally one in a million (<7,000 in 7,000,000,000). And the world could *really use* your unique set of skills and talents.

What is wanted and needed in this world right now is leadership and full self-expression. We need you to be unabashedly who You are and lead the way. I am requesting that You step up and play the role that You came to this Earth to play. There is no one else on the planet who can fill Your shoes and do what You are here to do. Only You. If You don't do it, it doesn't get done. That is why we need You. That is why You are so critical.

Be Like a Cat

NASA Astronaut Mae Jemison, who was the first African American woman in space, was speaking to a small group of female space engineers at Virgin Orbit about impostor syndrome —that feeling like you don't belong or are not smart enough or important enough to be where you are.

Her advice to us was, "Be like a cat."

The room filled with quizzical looks… She continued, "Have you ever seen a cat that didn't think that it had every right to be right exactly where it was?"

I thought that was such a great answer I had to pass it on. So next time you are doubting yourself or feeling out of your element, just remember to channel your inner feline.

You Can't Observe a System Without Affecting It

You are affecting the world all day, every day. Think of yourself as your own little media empire. You are broadcasting facts, emotions, responses, energy, and emails all day. Every snide comment, every detracting word, even every negative thought has an impact. It is not harmless. It is actually altering the future and creating an environment that makes it harder for the people around you to succeed and even for you to feel happy and fulfilled.

Sighing in a meeting, taking too long to respond to a question, looking away because you are annoyed, avoiding someone on your team—all those things are noticed. They erode team trust, team effectiveness and your own satisfaction. It makes a difference. A Jedi, or anyone who wants to make a difference in the world, has to hold themselves to a higher standard.

If you don't like the way someone is acting, instead of stewing about it or pretending you don't care, go apologize *to them* generously, sincerely: "Sorry Marc for not giving you my full attention in the meeting. I let myself get too worked up about how some of the decisions were being made. I thought I would come and talk to you about it instead of stewing and see if there is anything I can do to help us find a solutions that will really work for everyone."

Even how you walk into the room for a meeting matters. Are you ready to contribute and move the ball forward or are you sucking the air out of the room? Aim to be someone who is helping things move forward. Bring some energy to things.

It's not your manager's job to make the meeting work, it's yours. True leadership means doing what it takes to make things work no matter what your job title is.

What Is Your Challenge?

When I went to summer camp as a girl, the first night of camp we gave each other challenges: one girl would be challenged to listen better, I would be challenged to include others and to give other people a chance to lead, someone else would be challenged to be more considerate, another to speak up more. When I meet people from camp now, I can just say, "What is your challenge?" and they can tell me what they are working on in their personal training and development. It's a habit, way-of-life that continues.

It's a habit all Jedi can adopt, to know what you are working on. You can always be working on your edge, always looking for where you can improve, be more generous, more patient, more inclusive, more diligent, more fun-loving, more open.

So what is your challenge? What are you working on? The reason we ask is so you have someone around you supporting you. So be sure to share what you are working on with trusted friends and fellow Jedi. They may even have something to offer you on your path.

Listen for the Gold

Give people your full attention. Whether you are in a meeting, or a class, or out with friends, how YOU listen matters. The energy you devote to listening to them is felt and actually can improve their performance, their ability to articulate themselves, even their ability to make connections and come up with new ideas. It is like giving everyone around you the "home field advantage" of having thousands of fans cheering them on.

You are that powerful. You can make that big a difference.

Conversely, you can also cause a speaker to lose their place, or a manager to stumble on their words if you are being rude enough. You, just being checked out, can be a disruption, an energy suck. All of us have been around people like this. You don't want them there. It's hard to even think when they are around.

So cultivate your ability to listen powerfully and you will start to bring out the best in everyone around you.

Practice this week giving people your full attention and beaming at them when they talk, be interested, ask questions, be encouraging. Notice the impact pouring so much attention into a conversation has on you and on them.

It's Not About The Nail

There is a great two-minute online video called "It's Not About the Nail" that you should search for and watch. It's a great reminder that sometimes the best thing we can do for someone is to just listen to them quietly until they feel *totally* heard.

When my husband talks about something going on at work, my first instinct is to back seat drive and tell him what I think he should do! But that doesn't give him a chance to clear his mind, to get everything off his chest and he is usually not ready to hear any input yet.

But if I hang back and just listen and listen and keep listening and saying, "what else are you concerned about?" Or "Why does that worry you?" and letting him talk more and more he starts to feel lighter, like a weight is being lifted.

Then once his head is a little clearer and he realizes he has been the only one talking this whole time he will usually say something like, "What do you think?" or "What would you do?" Remember people are more likely to listen to your input if they asked for it.

So wait for them to ask; generously, lovingly listen until they ask.

Speaking with Respect

If the way you *listen* to people impacts their performance, imagine how much room there is for improvement in the way you *talk* to people.

Say, "Alex, I love the new design you worked up. I have a concern about the passive cooling system. Did you look at the heating loads on orbit yet? I can help with that if you need." A lot of it is tone of voice and body language. A Jedi has to be very mindful not just of the words they use, but how they are heard and felt by the listener. If your listener gets defensive or upset you can address it directly, "I feel like I upset you with how I said that. I'm sorry, I really do want to help. Can you suggest a better way to say it?" The tone with which you say that matters, too.

At the other extreme, I could say coldly, with a quizzical look, "Hey, I wasn't trying to be critical, I was just trying to help." You can see how that tone is not likely to create a lot of warmth and trust between the parties.

If you have had one or more interactions with someone that didn't go well, you will likely stop talking to that person as much and subconsciously or consciously avoid them. I assure you that won't help us get to space. Your job as a Jedi is to work to get back into the highest possible groove with as much of your team (and beyond!) as possible. Consider also that for the purpose of human space flight, there are thousands of people on your team, not just the folks in your chain of command!

You're a Sophisticated Tone and Mood Sensor

You can *feel* the difference between someone speaking to you in a condescending tone and someone speaking to you in a respectful tone. You are a highly sophisticated tone and body language sensor. The best supercomputers in the world cannot yet pick up on the things that you can detect without even trying.

In Benjamin Zander's brilliant TED talk about classical music he says, "If your mother calls… and says, 'Hello,' you not only know who it is, you know what mood she's in! You have a fantastic ear."

Train yourself to pay attention to your sensors. Listen to them.

Sometimes our brain will pop up a dialog box in our head that says, "I think he is annoyed" and we will just dismiss it without really paying much attention. But if you really stop and not ignore any of those warning signs, you can help co-workers work through frustrations and upsets before they boil over into something that can impact them or worse yet, the team and the project.

If you aren't as good at reading people, or someone is hard to read, feel free to calibrate your sensor. Check in with someone after the meeting, "Did my questions offend you?" or "Were you getting frustrated with Alan?" Sometimes someone just giving voice to the things that are unspoken can help relieve tensions and give people safe outlets for their stress.

If they are offended or frustrated, you can apologize or offer to listen. In either case you will both return to your desks in a better headspace to get back to work.

Speaking With Love

I got a good lesson in the importance of nuance and tone when I moved in with my husband George. I call that period of our relationship the "Taming of the Shrew" period.

I was a fiery-mouthed Latina and he was an even-keeled New Englander. I would find a dish in the sink and I would say, "Stop leaving your dishes in the sink!" and he would say, "Hey, please stop yelling." To which I would laugh, "Oh, *that* wasn't yelling." But what he taught me from those many exchanges is to be painfully mindful of my tone of voice.

Amazingly, I couldn't even HEAR what he was objecting to when this started. I first had to train my ears to pick up the condescension and derision in my voice that he was referring to. Over time I became more and more sensitive to it and I learned that I often spoke in a tone that was dripping with contempt for other people—not a great way to build people up!

I want you to train your ears to pick it up, too. Listen for how often you are condescending in your tone of voice or your words.

Experts say that when you use a critical tone to correct someone, they DON'T EVEN HEAR WHAT YOU SAY. (Which is why mothers complain about having to say things 1000 times). So I learned painfully, over time, to walk into the kitchen, see the dish, feel the heat rise up in my throat, take a breath and then say, "Hey honey, next time can you put the dish in the dishwasher?" as lovingly as I could muster. I wasn't being fake, I wasn't being saccharine; I was mindfully choosing a tone that matched the result I wanted to produce. Amazingly, it worked and we survived an important challenge to our relationship.

Instead of snapping, "What the hell were you thinking?" try calmly saying, "I was upset that you threw out my favorite t-shirt. I know it had some holes, but it had sentimental value for me. Next time can you ask me first?" You'll be amazed how much more effective calmly sharing your perspective is.

Take-home point: If you want someone to change a behavior, speak to them with love and respect. If you speak to them in frustration or annoyance, it will be less effective, or if it does produce the result, it will cost trust and affinity in your relationship.

Be a Jedi Audience or Meeting Attendee

- If there are empty seats near the front of a talk, fill them in. It supports the speaker to have a full and attentive audience.
- If there are empty seats at the table at a meeting, sit in them instead of around the edge of the room (unless there is one seat left and your boss is still coming).
- Make eye contact and give the speaker your full attention.

- Ask respectful questions that support fulfilling the objective of the meeting. Speak as if you are addressing your favorite person.
- Stay off your phone and computer. If you are taking notes or covering the event for social media make sure people know you are "on task," not doing something unrelated.
- Make sure not to dishonor anyone in the room (or out of the room).
- Always raise concerns (and be open to the fact that there may be factors you are not aware of).

Never Say Never

Alarm bells should go off whenever you use the words "always" or "never" (unless it's a safety rule!). Usually when we are using "always" or "never" to describe a person, we are making a sweeping (and damaging) generalization. "He always cuts people off" or "She is always late." If we lock people into a box like that, we make it all the harder for them to appear any differently to us than the way we expect.

Leave the future unwritten. Leave room for them to show up differently. Say, "She has been late a lot in the past." Or "Sometimes he cuts people off, but he is working on it." How you say things matters.

Jedi Patience

Recently, I saw how a lot of the anger I express toward my kids comes from having no patience. As an aspiring Jedi Master, this was quite a blow. I could just hear Yoda in the background saying, "I cannot teach her. The woman has no patience."

I decided to dedicate myself to developing patience.

How do I practice patience? It takes giving up being attached to getting my kids teeth brushed by 8 p.m. It's a calmness, a surrender to 'this is what it is going to take to be successful long-term,' a willingness to go the long way, to skip short-cuts, to do things as they are meant to be done. Not as a punishment or an added hassle put there just to annoy you or slow things down.

Practice looking at challenges in your life as an opportunity to engage another person in your effort or idea. "You haven't seen our test bed? Come check it out; it's amazing." Maybe that extra approvals step will not only win you a new ally and support, but it could also get you excited again about your project by showing it off. It could also remind you of the big picture and help you to be more mindful. It could also spark new ideas, like: "I should invite the manufacturing lead over here, too. Maybe she'll have some useful insights for me as well."

Patience doesn't mean you will go slower. You might finish a project faster without the energy wasted on getting angry and frustrated with yourself or your code or with what some other team did that you didn't like.

A Jedi Craves Not These Things

"This one a long time have I watched. All his life has he looked away...to the future, to the horizon. Never his mind on where he was. What he was doing. Hmph. Adventure. Heh. Excitement. Heh. A Jedi craves not these things."

—Yoda

For years I have resisted the idea that a Jedi does not crave adventure and excitement. Obviously, all the Jedi that I know do crave adventure and excitement. What do we do about that?

What Yoda is saying is that there is work to do. That we are not training to be Jedi so that we can have epic laser battles in an X-Wing fighter. We have to be completely dedicated to the cause of maintaining peace in the galaxy and doing whatever work there is to do in support of that. If that means not seeking out an "exciting" battle, then we won't. If you are in it for the excitement and adventure, then you won't listen to the Force when it guides you to stay home and help plant the crops or teach what you have learned.

Why be a Jedi then?

I certainly have always wanted adventure and excitement. I have even had a lot (going to the Arctic and the bottom of the ocean, floating weightless, working with some of the most influential people on the planet).

In my experience, the flashy things have never been as fulfilling as I think they will be. Afterward you still feel like the same person. I think that is what drives so many rich/successful/famous people crazy. There is no there, there. There really is nothing at the top. You go to a movie premier or a fancy dinner with A-listers and you still go home to your same life.

Real joy and fulfillment comes with being present, with loving who you are and with helping and connecting with others.

Adventure and excitement are pretty empty without that.

Building the Culture to Go Along With the Spaceships

I have talked about wanting to create the *culture* of *Star Wars* and *Star Trek*, not just the spaceships. But what does that take?

One piece of the puzzle would be to build a city in space as a model that all cities on Earth can look to. They might say, "Wow, it is possible to avoid strife and violence, it is possible to have a minimal ecological footprint, it is possible to take care of all people and for all people to be powerfully engaged in their lives and their missions!" That "bootstrap" method of helping all cities see how they can pull themselves up has always appealed to me.

Lately, though, I have realized that before we can build that city in space we also have to make our individual lives a model of what is possible. For example, I need to do the work to get my own life to work, to live transparently, good and bad, and to be a model for how life can go. If we want to create an awesome city that works, we first have to figure out how to create an awesome life that works. I have to figure out how to make my marriage work, how to work powerfully with anyone, how not to scream at my kids, how to live healthily, how to stay connected to friends and family, how to make a difference for others, and how to stay true to what I came to Earth to do.

So my focus has been a lot less on huge projects that can impact a billion people and a lot more on what I need to do to get my own life to work, who I need to be, how I do that work.

I am reading more than I have ever read before, I am calling and meeting and connecting with more people than ever before and I'm focusing on consistency and follow-through more than I ever have before. I have opened myself up to coaches, professionals, and friends for their contributions more than I ever have before.

Before now, I would have thought those things were small, personal goals, but now I see them all as critical prerequisites to any work there is to do to contribute to the world and to be able to stand firmly for a future where all people experience being alive as a great gift.

So I am starting with me. And my invitation is for you to start with you.

If enough of us are on that path and in sync with that vision for humanity's future, we will be able to create it. From there we can do anything.

It Takes A Village

For me to stay true to what I came to Earth to do, I need a community around me. It is like having my own Jedi Council. When I don't feel powerful, I call one of them as quickly as I can. They remind me of who I am and what I stand for, even when I can't see it myself. I do the same thing for them. I cannot do it alone, nor can they. It takes all of us standing together to pull each other through.

There is an African Proverb that says, "If you want to go fast, go alone, if you want to go far, go together."

So look around you. Who do you see who would be up for playing that game with you? Who wants their life to make a powerful difference? Who else do you know who always wanted to be a Jedi? Maybe one of the things you ask for is people to be on your personal Board of Directors, or your Jedi Council, or to start an accountability circle with you, or to start a leadership book club where you get together once a month to talk about leadership books you read.

It's time to build your team. People you can trust. People who will listen to you as awesome no matter what and hold you to it. People who are positive. They say you are the average of the five people you spend the most time with. Be sure to stack your team with awesome people!

Band together and let's change the world.

Connecting With The Force

"Luminous beings are we, not this crude matter" –Yoda.

Which character from Star Wars are you most like? Which character would you most want on your Mars Crew? I am most like Princess Leia, but who I would most like on my Mars Crew is Yoda. He is calm, wise, and very powerful.

What can we all do to be more Jedi-like?

Jedi Masters sit still, relax their thoughts, and feel the Force flow through them. You can practice this too for a focused amount of time every day. You can do five minutes to start, then work up to 15 minutes, then up to 20 minutes or more connecting up with the force that flows through all things. It is a great meditation practice and a great way to calm down, to center your thoughts and to reconnect with your purpose.

It is like charging your phone. You wouldn't think of asking your phone to go a day without being charged, so do not ask that of your body, your heart, or your mind. As you sit to meditate, you can even imagine yourself plugged into the universe, getting supercharged with energy, with ideas, with inspiration for the day.

Meditation will become a part of our culture the way coffee is to us now. Many people wouldn't dream of starting their day before their morning coffee. Just think, it used to be that way with a morning cigarette; many people would reach for their pack before even getting out of bed.

We ask our grandparents, "You guys smoked two packs a day *indoors.* What were you thinking?" Our grandkids might ask us instead, "But how did you think clearly without meditating every day?" Meditation will grow in acceptance the way yoga has. From something virtually unknown in the U.S. in the 1960's to now seeing five or six yoga studios just here in Lancaster, California (where we live).

Mindfulness and a daily meditation practice will become a normal part of our daily routine.

As a Jedi, it is an important opportunity to connect with the Force that flows through all things. It connects us, binds us. Life creates it, makes it grow. Connecting in with that life force daily will help you stay grounded, present and receptive.

Keeping The Big Picture Present

What does it take, in reality, in day-to-day life, to be someone creating an inspiring future for humanity?

Commitment, dedication to bettering yourself, finding like-minded people to hold you accountable, and exposing yourself to new ideas.

It also takes keeping yourself present to all of these things every day so that you are choosing to respond thoughtfully and with your long-term commitments in mind, rather than letting yourself react knee-jerk as things come at you.

Look at your daily routines, who you talk to regularly, what you read, how you invest your time. Look for ways to keep the big picture present for yourself. Read your Mission everyday, keep it where you can see it. Take a hike or some perspective-expanding experience every week. Watch the sunset. Call old friends. Have an end of year ritual with your friends to review the year and set goals for the coming year.

As my husband likes to say, "All your problems look small from Low Earth Orbit." Keeping perspective will help you keep going, even through the tough times. When in doubt, zoom out. You're okay, you are going to make it.

Here are a few more suggestions for building a lifestyle that supports you, your mind and your spirit.

Read a Book

I didn't read a book for many years. I was "too busy" to read books. Then I noticed that I can usually find time to read Twitter or Facebook for a little before bed. I realized that if I used my social media time to read books I could actually finish a book! I also realized that a good book was much more enriching and inspiring than social media.

Read something that is challenging for you, something that will help you grow.

Listening to books is even better, because I can find even more time to do that. While driving, while folding laundry, while doing the dishes, or while at the gym I can be listening to a book. Most books are about 6 hours long and with the amount of driving and household chores we have, you can easily do a book a week or a book a month. You can even get them from the library for free!

There are the classics that I thoroughly enjoyed reading like *Illusions: Adventures of a Reluctant Messiah* by bi-plane pilot Richard Bach, *How to Win Friends and Influence People* by Dale Carnegie, *The Alchemist* and *Warrior of the Light* by Paulo Coelho, and Viktor Frankl's *Man's Search for Meaning*, about his time as a prisoner in the concentration camps of Nazi Germany.

Recently I have also enjoyed *The War of Art*, by Steven Pressfield, which is a fantastic (and easy to read!) guide to getting up and doing your calling everyday in spite of everything pulling against you.

It doesn't all have to be leadership books and classics. I have found that good fiction or biography can help me grow as well. I can see myself in characters in *Where'd You Go Bernadette?* or be inspired by the true story in *Unbroken* or called forth by the powerful dystopian future in *Parable of the Sower* by Octavia Butler.

What books have inspired you in your life? What characters have called you to grow?

What are you putting on your reading list?

Finding a Coach

Many people often recommend finding a mentor. I would say: great, mentors are good at giving advice and sharing their experience, and ideally even becoming advocates for you among management.

I also recommend you find a coach, someone who is not afraid to tell it like it is. "Your jump shot stinks." Someone who will tell you what you need to do to get your game to where it needs to be. "Stop eating cheeseburgers, be in the gym every morning by 7, do a double workout, and don't leave until you make 12 jump shots in a row."

A coach is someone who sees your potential and is more concerned with your reaching that potential than you liking them. You may need to hire a coach, but the plus side is that they will have more bandwidth for you and that if you are really serious about being great, that is what it takes.

Anyone who is serious about being the best has a coach. Performers have coaches, athletes have coaches, executives have coaches, so if you are up to something, consider getting a coach. You are worth it.

Ride the Horse the Direction It is Going

If you get on a horse facing backwards, you will have a heck of a time getting where you want to go. You may feel powerless, along for the ride, at the mercy of wherever the horse takes you.

If instead you get on the horse facing forwards, and *ride it in the direction it is going,* then you will have a better chance of steering it and shaping where it goes. You become the driver in your life.

Instead of resisting what is happening, go along with it. "Great, let's all go to Disneyland with your parents. Sounds like fun. I'll pack the Mickey ears. Can we make sure our hotel room isn't next to your parents so we have a little more autonomy? I am happy to do the bulk of the driving too if you like." (To make sure we can stop at Starbucks when I need.)

Daily Action

"The only thing that makes a difference is action."

To really dive into your training, I invite you to commit to a daily action and do it every day, no matter what. Now that you have accepted the call to be on this journey it is time to step into the Special World. We will do this by interrupting the ordinary auto-pilot of life and choosing everyday to do a daily action of our own choosing, an action that will help wake you up to the bigger goals you are committed to. This is the next step on the Hero's Journey.

When you miss a day, don't use that as license to be harder on yourself, just recommit and start again and try to learn from it.

Pick one action that you will commit to doing every day. It can be whatever you want, just pick something that will interrupt your auto-pilot way of operating and require you to actively attend to completing your action every day.

You might even already know what you need to commit to. It will likely be what you are most avoiding or resisting committing to do every day. It should be mildly intimidating, yet something you know will produce a result you want or have you become who you want to be. Don't be afraid to swing out. Once you are out of school, it is easy for your growth and development to stagnate if you don't keep pushing yourselves out of your comfort zone.

If you don't have a deadline to read the book by, or a coach telling you to run another lap, it's harder to do it. So set up structures that work for you. Make your own deadlines and get people who will hold you to them. Make it a ritual. Do it at a set time every day. That will make it easier for you to succeed. Set reminders on your phone! Give yourself a reward if you do it 5 days in a row.

Be the person you have always wanted to be today, not someday.

What are you going to take on doing every day? Make sure you tell someone and give them permission to hold you accountable for completing it.

It starts today.

Review

You cannot observe a system without affecting it. How you are *being* matters. Watch out for:

1) The energy you are putting into the room
2) How attentively and respectfully you listen
3) The respectful language and tone of voice you use when speaking

Homework:

1). Commit to one new action that you will take every day. Pick something that will get you off autopilot and in action creating the life you want. Tell your buddy. Set your phone to remind you!

2). Practice listening with your whole being. Listen to others as if what they are saying is the most important thing you have ever heard. See what opens up.

3). Check in with your buddy. Support each other in your daily actions, your end of training goal and using the material from this section.

JON MARRO

CHAPTER 3

The Dark Side

This journey is about learning to bring balance to the Force. Balancing out your inner dark side with your light.

As I have said, I am holding you to a higher standard—especially if you are a part of the space community. I am committed that our industry be the vanguard of a new future for humanity, not just through our high-tech gizmos, but in who we are being, how we are treating each other, how we speak. Just as importantly, I want us to do this because it will make us more effective, more likely to get to Mars in our lifetimes, more likely to lower the cost of access to space. We need to be operating at an extraordinary level to achieve extraordinary results.

I am passionate that when we go into space, we take the best aspects of our humanity with us. That we use the move to space as a chance to grow as individuals, to grow as a species, to acknowledge our faults, to own our weaknesses, and to consistently choose not to let them run the show for the future of our species, our home planet, and of space.

If we can use the transition to becoming a space-faring civilization as a catalyst for personal and societal transformation, then we will have done something that is really game changing. And that is something worth facing all your inner demons to accomplish.

Facing Our Dark Side

We all have a dark side.

Whether is it being lazy, jealous, scared, angry, judgmental, greedy, arrogant, meek, or addicted to food, drugs, alcohol, shopping, video games, the internet or sex. Some of us experience debilitating bouts of depression. Some of us are reckless, or condescending, or don't follow through for ourselves or others. Some leave people always feeling less than, some let their negativity and sour mood affect anyone who comes near them. Some have Impostor Syndrome where they feel like they don't measure up and live in fear of being found out as a fraud. (I did a talk about my impostor syndrome at a TEDx conference).

Whatever flavor your dark side comes in, it gets in the way of your mission, your relationships with others—and yourself. People either don't like being around you because of it or no matter how close they get, they never know the real you because you are hiding it. Or, you surround yourself with people like you and agree not call each other on your garbage.

On each of those paths though, in the end, your true self is left alone and unfulfilled.

In *The Empire Strikes Back*, Luke is training in Dagobah by running through the swamp while listening to Yoda. Yoda says three very important things in this scene:

1) "Beware of the dark side. Anger, fear, aggression; the dark side of the Force are they. Easily they flow, quick to join you in a fight. If once you start down the dark path, forever will it dominate your destiny, consume you it will."

2) Then Luke asks, "Is the dark side more powerful?" "No, no. Quicker, easier, more seductive."

3) And then lastly when they get to the foreboding cave, Yoda says, "A domain of evil it is." "What's in there?" Luke asks. "Only what you take with you," Yoda replies.

Going into the cave is about facing the dark side within you.

Part of your Hero's Journey is to confront what you already know is there and also what is still in the shadows unknown to you, or known but stuffed down so far that you hope it never resurfaces.

The reality is that you cannot stuff things down that far.

You have to face your worst, your most hurtful, selfish, cruel, embarrassing, secret, and shameful stuff. This is important because it hangs over you, stealing your energy and vitality and keeping you separate from other people as you try to keep your secrets hidden.

The best antidote to dark is light. They say sunshine is the greatest disinfectant. Being willing to own your stuff and tell the truth, at least to yourself, is an important first step on your journey.

Another benefit of owning up to the darkest aspects of ourselves is that it can give you a great deal of compassion for others' struggles. If you can accept all parts of yourself, then you can accept all parts of others. If you reject parts of yourself you will reject that in others too.

Keep in mind that to accept your dark side doesn't mean you like it or you approve of it. It just means that you accept that it is there. You allow it space to be. You loosen its grip on you.

Once you can be with all parts of yourself, then you can connect with anyone, no matter how dark, and know they are redeemable (just like Luke's father).

WRITING EXERCISE:

Turn to a fresh page of your journal. Write the date and "Dark Side" at the top of it and list out some of the darkest and most sinister aspects of yourself, your darkest moments, the things you are least proud of, that you would rather not be a part of you. You don't have to incriminate yourself, you can write it in a way that only you would understand, should this journal fall into the wrong hands.

Going Inside the Cave

Just as someone in Alcoholics Anonymous can say "My name is John and I am an alcoholic." To be a leader it is important that you are able to stand up in public and own anything about yourself.

People can handle just about anything if you are honest and at peace with it yourself. So doing the work to make peace with all parts of yourself and all aspects of your past is an important part of your training.

How does your dark side interfere with your mission? How does it affect the people and things that you love?

My goal for this section is to have you go into that cave, see the darkest aspects of yourself and to come out knowing that they do not define you or limit you, that in fact they make you human and even loveable.

It is through our imperfections that we can connect to others. I see it as the chink in my armor that allows people to see the real me. It is my humanity, my vulnerability that lets people in.

It's a feature not a bug.

Here is what I want you to explore for the next two weeks: Just letting it be. Not trying to fix it or change it or hide it, just sit with it and see if it loses some of its power.

The Way of the Jedi Warrior

“What the warrior renounces is anything in his experience that is a barrier between himself and others.”*

Renouncing anything that is a barrier between me and another person calls to me and demands of me a higher level of being.

It reminds me to let go of all the judgments I can make in a moment about anyone and anything: “He can’t remember where he parked the car,” “she forgot his name,” “he still has not done what he promised he would do.” Subconsciously, I tend to “keep score” in order to win or come out better than others.

It is something that I know I do, but didn’t see any way to stop. Now I have something more important to me than being better than everyone. I'm giving up anything in my experience that is a barrier between me and another person. I am choosing to be connected to all things and be a Jedi Warrior.

After all, if the Force is flowing through all things, then to be connected with all people and to remove all barriers between them and me would leave me closer to the Force.

I will leave you with that to meditate on today. *What the warrior renounces is anything in their experience that is a barrier between themself and others.* What are you willing to give up in order to achieve your true potential?

So now we will begin to delve into the Dark Side. The things that keep us separate from each other and from the Force.

*From *Shambhala: The Sacred Path of the Warrior* by Chogyam Trungpa

Don't Throw Anyone Under The Bus

In 2013, SpaceX and Blue Origin were in a dispute over who would have the use of NASA's retired Space Shuttle Launch Pad 39A. NASA had awarded the exclusive use of the pad to SpaceX, and Blue Origin contested the decision with the Government Accountability Office. During this time, SpaceX's CEO Elon Musk made colorful comments about Blue Origin, questioning the likelihood of their using the pad for orbital launches in the relevant time period. It felt like Elon had thrown Blue Origin under the bus.

I don't think Musk would have made those comments if he related to Blue as a critical part of humanity's effort to get humans to Mars, something that is very important to Elon and to many of the people at SpaceX. Yet, Blue Origin is important to that future.

Perhaps an employee now at Blue Origin will move to SpaceX to work on the Mars Program (but will be a little tempered because of that comment), and perhaps Blue would help lobby the government along with SpaceX to make sure our regulatory system supports private humans-to-Mars efforts. Perhaps Blue will make one of the systems that make it possible, or help raise the capital required.

This industry is just too small for us not to support each other. We are in this together, for the long haul. I want all of us to succeed: NASA, SpaceX, Blue Origin, Virgin Galactic, Orbital ATK, Sierra

Nevada, Masten, Northrop, Lockheed, Boeing, ULA, and the many new space enterprises starting up.

Otherwise we are just going to go into space and take all of our problems with us—same species, different planet. I think we can do better than that.

Now here is the thing: shortly after Elon's comments I realized *I* have been throwing Elon under the bus. People I meet would sing his praises and I would say, "Yes, he is great *but*…" and then try to undermine him and talk about the things he could be doing better!

Easy for me to say. I don't have multiple industry-changing companies to run and thousands of employees, stockholders, and customers to manage! I also realized I wasn't walking my own talk, I wasn't being the change I wanted to see in the world. So I stopped.

What most of us don't realize is how often we throw people under the bus. We are so attached to our idea or viewpoint that it highjacks our brain, and protecting that viewpoint becomes more important than getting to the stars, more important than being fulfilled, more important than having a team that works.

Stop and think about that. We let that view become more important than the most important things in the world to us.

We even throw the people we are closest to and love the most under the bus! My opinion about my husband's faults can, in about a nanosecond, become more important than being happy, more important than raising great kids, more important than feeling loved and connected. Not because I intellectually agree with that, but because when it comes down to it, I still complain about his leaving the lights on all the time (I mean really, doesn't he know how much energy he is wasting?) in spite of the fact that my complaining degrades all of those things that matter to me.

Humans are funny creatures.

One of your assignments this week is to notice how much you throw people under the bus both at work and in your personal life.

Transcend 'Us Versus Them' Thinking

Wisdom is expanding your view of who "us" is from your family to your team, to your company, to your industry, to your species, to all of existence.

Sometimes it feels like we need a "them" to compete against, to fight, in order to get our best performance. Some people even wish an alien species would invade as a way to help unify all of humanity.

But what if we didn't need to fight each other (or even aliens)? What if we turned all that energy to fighting things we all struggle with, like the gravity well or entropy? Instead of competing with a different rocket company, couldn't we align on our common passion to "break the surly bonds of Earth"?

'Us versus Them' thinking is very limited and old school. If we really want to become an interstellar species, we will need to trade in Us versus Them thinking for some bigger picture thinking.

The Screen Saver of Your Mind

The thoughts that run through my head can get pretty nasty. So I am upping the bar on myself to not let myself even *think* things that I wouldn't say. My practice has been to replace the virulent thought with one I endorse. From, "Leave me the F@$% alone!" to "I need a little space right now."

I have been inspired by UCLA Coach John Wooden's rule: "Make every day your masterpiece." I have been asking myself, "Is thinking that way going to make today a masterpiece?"

My thoughts are best when I am reading a lot of wise words and quotes. I think of them as the screen saver of my mind. When I am reading Dale Carnegie or quotes off my Chill app, those get stored as the screen saver of my mind. Those thoughts influence and color my thinking, my reacting (or non-reacting!) and my day.

So upgrade your thoughts. Pick a brain/mind screen saver you like and get rid of the default one (if it is negative like mine). Actively cultivate regular exposure to ideas that uplift you. Limit exposure to ideas that don't inspire and enliven you.

Don't Type Things You Don't Want Forwarded

If you are undermining people, stop. It won't help us get to space any faster.

We say things like, "Well, Bill doesn't know what the hell he is talking about." Or "He likes to think he is a systems engineer, but he isn't. Don't listen to him."

If you have a legitimate concern about someone, say it in a way (and in a tone) that you wouldn't mind them hearing, like "Bill's training is in electrical engineering, I have concerns about our basing structural decisions just on his input. I'd like to hear what Kyle and Melanie think too."

Next time you are about to open your mouth or press send on that email, ask yourself, would I mind if this was made public? Is this how I want to be known in the world? Strive to have your words be consistent with what you say you are committed to. It is a high bar, and a great check to put on your speaking and writing.

Giving Up Sarcasm

Another way we throw people under the bus is with sarcasm, "Yeah, like we haven't tried that seven times already."

I know, I grew up with sarcasm as my native tongue. In my family it was almost an art form. It wasn't until I got into college that I realized how caustic I was. I went cold turkey and gave it up entirely because I realized it was so often at someone else's expense.

Humor is great. It can lighten the mood, ease tensions and build comradery. Snark and sarcasm can lighten the mood and ease tensions for some but often at the expense of someone on the team, or at the expense of another team. Often it can end up being a net-negative for space as that distance and mistrust can lead to breakdowns in communications that lead to actual in-flight failures.

Start listening to your sarcasm and if you hear it is at the expense of someone else, see if you can find another way to be funny!

Stress

When I was in graduate school I really struggled under all the pressure I had put on myself to save the world. The stress was debilitating.

I was drowning in all the problems the world needed to solve and felt stressed out, overwhelmed, and unsure where to even start. It all just made me feel like more of a worthless failure.

One weekend, though, I was taking a personal development class and I suddenly got freed up from all the stress I was putting on myself. I realized that the world didn't need saving. It would be fine with or without me. I don't HAVE to do save it. AND I could choose to take on the problems I was most interesting in, not because I HAD to, but because I WANTED to. Because what more interesting thing could I possibly do with the next 70 years? What would be more engaging, satisfying, challenging, entertaining than taking on a huge epic quest to change the world?

I still want to alter the trajectory of humanity, now I am just doing it for fun and for a love of the game instead of because I have to. Somehow that shift made a *huge* difference for me. I didn't get as bogged down in the significance of everything—which actually made me a ton more effective!

So look at your big dream, your inspiring mission, and take a deep breath. You don't have to take yourself so seriously or make it really stressful and significant to be successful. All you have to do is stay light and keep some perspective about it. Remember you are doing it just because you said so (sometimes I even like to think of life as just a REALLY impressively high-fidelity video game that I play just for fun). Keep working at it every day, one foot in front of the other, honoring your word.

What can you do to enjoy your life, your goals and the path you set out on to achieve them without the stress?

Dissatisfaction

The beginning of a new job or a new relationship is often filled with joy, excitement and fun. We throw ourselves completely in and are willing to do whatever it takes to have this be the job or relationship of our dreams. We show up early, stay late, buy our partner flowers and beam at them anytime they walk into the room.

It is called the honeymoon period.

At some point though, things usually start to change. We are a little less enthusiastic. We are quicker to see faults in management or the love of our life. Over time this can erode our entire experience of being alive and we start to think that maybe this isn't the dream job or relationship after all. Maybe we should leave.

My friend Anurag says that usually the only thing that changes during that period is where we are putting our energy. In the beginning we are willing to *invest* all our energy into the job or relationship. We are putting tons of love and attention and effort into it and in fact just the simple act of putting tons of energy into something can make it great.

When things usually turn to dissatisfaction is when we start to coast, start to expect something back, start to think, "Wow, I have been putting so much energy into this, I should really be getting more back;" "I should have had a raise by now;" "She should not be nagging me nearly as often as she does."

If you ask people to map when their dissatisfaction with a start-up, a marriage or any other project began, it usually starts about the time when they stopped putting energy in and starting wanting to get something back.

If you are dissatisfied with anything, try going back into honeymoon mode. Give it everything you've got. All the love and attention of someone who just fell in love, who just arrived—and see the difference it can make in your experience. They tell people thinking about divorce to stop talking about the problems and start doing they things you love to do together. Have fun again. Focus on what had you fall in love in the first place.

Chances are the flaws they have were there then too, only in the beginning they were charming and endearing, now they are insufferable. Sometimes, if you can get back the love and the connection, you can start to see their faults as endearing again.

The same is true at work. It's up to all of us to keep our Dream Job feeling like a Dream Job! If it's not, put more energy in, make this the place you want it to be, talk to management, offer to help, get in action to turn it around.

Same Boss, Different Company

Sometimes you do get a boss who is not supportive, who is frustrating and possibly even incompetent. Often it seems clear that the only thing to do is to change jobs. So you change, it goes well for a while and then again your boss turns work into a painful place to be.

You leave again, worrying that maybe the whole industry is broken. It happens a third time.

Perhaps it is not that bosses just *are* that way. Perhaps it is something in you that brings it out of them. That is why you keep running into the same boss at different companies.

Maybe the Universe put that person in your path to teach you something and if you run instead of grow, it will just find a way to put you back in that same situation until you learn the lesson that is being given to you.

You can even think of it like a video game. This person is just the dragon that you have to slay (or befriend) to beat Level 12. Even if you move or change jobs, you will still need to master this to be able to go to the next level of the game.

The good news is that if it has to do with you, then you can do something about it, without even having to change jobs! If you can learn what your triggers are and how to be responsible for those triggers, and if you can learn to speak directly, honestly and respectfully with your boss, maybe you can turn things around.

Who knows, you might even find out that you like them! Or even worse, that they like you.

Analysis Paralysis

Are you prone to over-thinking things? Sometimes action is the best remedy. Even wrong action will teach you something, and eliminate one of the decision options from your matrix. So, go, do, build! Start something, make it happen, commit to writing an abstract, giving a talk, attending a conference, engaging a new supplier, proposing a new solution or program to your boss, or even asking someone out on a date!

Talking it through with someone else is helpful, too, especially if you start by saying, “I am overthinking this way too much, I need someone to help me get out of this loop.”

Remember too, in the end you will regret more things you haven’t done than things that you have done.

An image I find helpful is of Tarzan swinging from vine to vine in the jungle. It doesn't matter too much if you don't know where you are going as long as you keep your momentum up. If you figure it out three vines from now you can change course and head that way, but if you stop moving forward you will not be able to get anywhere else.

The bottom line—take action. Commit. Dive in.

Anger

I struggle with anger so much that I feel both very qualified to address it and completely unqualified to address it.

The bottom line is that anger, fear and aggression lead to the dark side.

Anger brings out the worst in me. It creates distance between me and others. It hurts my kids. It destroys my productivity and effectiveness and leaves me with a sort of anger-hangover, where I feel disgusted with myself and unable to give anyone much of anything.

The things that I have found help me with my temper are:

- Getting enough sleep
- Meditating every day
- Reading good books that give me tools and perspective.

I like reading books that help. The latest tool they give me usually works for at least awhile and just thinking about it all the time and hearing examples of other ways people handle difficult situations reminds me to get more creative in my responses!

So far I have not found a silver bullet. I do like to stop myself mid-rage and say in a completely normal tone of voice, "Oh, was I yelling again? Oh man there goes my gold star for today!" And turn it into a bit of a game. That at least signals to my kids that sanity has returned to the kingdom and they don't have to run screaming for the hills.

I also really like the Plato quote, "There are two things a person should never be angry at, what they can help, and what they can not." There is another one I saw posted in the window of a coffee shop, "For every 60 seconds you are angry, you lose a minute of happiness."

"It is he who is in the wrong who first gets angry." William Penn.

This is very true. Often I get the most angry when I know I should have been watching the children more closely, or should have started getting ready to leave earlier. When my integrity is even the slightest bit out, I am much more likely to fly off the handle at others. Now, I've learned to tell the kids, "Sorry, I am mad at myself because I knew I should have put away the milk and I didn't and now it's spilled."

There are people who will tell you that anger can be used for good. It can. But for people like me who can rage with anger, we cannot use it for good. USC Business School Professor Dave Logan told a story at NASA JPL about the only time he ever saw his father get angry. They were out at a restaurant and the manager was very rude to Dave's mother. His father spoke angrily to the restaurant owner and let him know that his behavior was not acceptable. As they walked out Dave asked him about the unusual outburst. His father was back to his normal, calm demeanor. He said simply, "Sometimes you need to use anger, but never let anger use you."

Fear

A friend recommended a book to me called, "Rejection Proof: How I Beat Fear and Became Invincible Through 100 Days of Rejection."

In it, the Author Jia Jiang recounts his adventures seeking out rejection by making unreasonable requests of strangers every day. In the end his struggles and pain and especially his success are a huge inspiration to any of us who have ever been held back from going for our dreams by our Dark Side of fear.

In the first part of this book I have a section entitled "Ask for Help" but the truth is even I can see where I get terrified at the idea of asking for anything big. Yes, I can ask for a letter of recommendation, or a tour, or help with an event, but asking for money, or people to do something that is hard or demanding, or asking for volunteers to do things for ME (as opposed to for a project or business) is still hard. I was inspired by Jia's journey to increase my own discomfort level on the things that I ask for.

They say 100% of the shots you don't take, never go in. So go for it, swing out, make an unreasonable request, and start asking people for things.

Many times the author found that even when people said no, they often had a good reason, or another thing they could help with or a recommendation of someone else to ask. Whatever happens will be more of an adventure than staying home and getting older.

Remember:

- The failure of living a life where you sold out on your dreams is much more painful than any 'no' could ever be.
- Being courageous does not mean that you don't have fear. It means that you have fear and you act anyway.

Aggression

Aggression is making other people smaller so that we can feel bigger. It is domineering, putting people down, stealing their ideas, and undermining them in the eyes of others. Aggression can also be physical or sexual, violating their personal space or their dignity.

Most of us are not bullies, but there are probably ways we all use aggression to make ourselves feel more important. Sometimes it can even be defensively, when we feel that we are being attacked.

Two small things you can do are:

1). Remind yourself that you are awesome and that you don't need to make others smaller to make yourself bigger.

2). Really work on putting yourself in their shoes. Think about life from their perspective until you can have compassion for them and want to build them up and not tear them down. Maybe they had a tough childhood, maybe they are really smart but not good at sharing ideas, maybe they feel inadequate.

Put your energy into building up the people around you and see what shifts. That is a much more interesting use of your power.

Arrogance

Although many of you may be accomplished, smart and well educated, it is important to avoid the pitfall of arrogance. The problem is that it drives people away from you. Space exploration (or whatever discipline you are working in) is a team sport. Do whatever you can to be welcoming of all people and to build them up.

It doesn't matter how smart you are, if you leave others feeling less than you, you will not be a net positive for the team. Check your ego at the door, be there to serve and more opportunities will open up to you.

Failure

In June 2006 I was fired from my contractor job at NASA Headquarters. I was called into the company owner's office and told to pack up my things and was escorted out of the building within the hour. I was so shocked, stunned and scared I didn't know what to do. I just sat in my car and cried.

I was numb for days. I felt like a failure. I had such big dreams. When I started at Headquarters I had thought that I could have started in the mailroom and I would have worked my way up to be the top. Now here I was fired. Fired for doing what I thought at the time was right, helping the Exploration Office get input from 18-35 year old on their new plans.

Now my eagerness to help was being seen as insider trading. Someone at headquarters helping one of the NASA centers get a contract to solicit the input over another Center. It was a landmine I didn't even know was there. I felt helpless and angry.

Then I got a call from General Pete Worden who has a PhD in astronomy, is a General in the Air Force and was the Director of the NASA Ames Research Center. He said to me, "Oh, I've been fired about six times. If you aren't fired you are probably not pushing hard enough!" It was the first time I had smiled in days.

Slowly, I began to recover. Anurag, a coach of mine, says that if you lost everything, you are resilient enough to re-build. Two years later you would probably be better off than you are now. He wanted to free us of the trap of being afraid of failure. To see that sometimes failure can be great, liberating and even leave you better off.

In retrospect, getting fired was a "getting kicked out of the nest" moment. It helped me to do something I had been too scared to do myself, go out and blaze my own trail, work on the projects of my own design. I started giving workshops on "Launching Your

Career in Space" at universities and blogging for Wired Science. It also taught me a very valuable lesson:

Sometimes failure is the best thing that can ever happen to us.

Criticism

In Dale Carnegie's book "How to Win Friends and Influence People" he talks about how President Lincoln learned the hard way how damaging criticism is to your relationships and to your desired outcome. Lincoln's scathing letter to a colleague, published in the town paper, did succeed in getting the town laughing, but it just made the subject dig in his heels and get defensive and resentful. He even challenged Lincoln to a duel (which in those days was a battle to the death to which you challenged your opponent when your pride was wounded). Luckily for us the duel was broken off before anyone died, but Lincoln learned a valuable lesson—that criticism is often not effective and not worth it.

Later, during the Civil War, when a general blatantly ignored Lincoln's orders to attack General Lee at a moment that probably could have ended the conflict two years earlier than it did, Lincoln wrote a calm letter expressing his disappointment at the lost opportunity, but in the end did not send even that. He knew that that would force his General's hand to get defensive, retire, or resent Lincoln. Instead Lincoln let the matter go.

There is a lot of research that shows that people respond better to the positive than to the negative. You may get people to comply with force, criticism, fear or aggression, but it will come at a price. That environment breeds resentment, mistrust, and apathy. Much more powerful is to do the work in yourself so that you can work with them to find a reason for themselves that would make them want to listen.

If you need the team to wear protective equipment, respectfully engage them on how the protective equipment was provided for their own health and safety and praise them for wearing it when they do comply. Ask them what they think the best way to stay safe on the job site is. You can also get in their world and see if the equipment is too uncomfortable or ill fitting; maybe if you can help them with that, they will wear it.

It's easy to say you can catch more flies with honey that you can with vinegar, but it takes real mastery to choose the high road when someone is not listening for the fourth time, or has just screwed up again (even after you have already talked about this!). That is when you really need to take a breath (or a walk) and remind yourself that this is an opportunity to train both you and them in skills that will help your project, your company and the world work. What works is to speak respectfully, and with love (and I use that word intentionally— I do not just mean 'be professional' which is office slang for 'thinly veiled anger and hostility'). I don't just want you to be professional. I want you to wait until you can literally come from a place of love, of respect, of seeing them as JUST LIKE YOU, a flawed human, doing their best, in need of a little guidance and open communication.

That is a very unusual conversation to have in the workplace. But it is one that might just break the pattern, and open the space for new healthy, productive workplace habits to emerge.

Jealousy

During one New Right Stuff Training I decided to skip covering "Welcome People with Competing Skillsets" during class. That afternoon Virgin Orbit invited Olympic Snowboarding Silver Medalist Gretchen Bleiler to come talk. I was a little jealous of her being invited to coach the company and not me.

I laughed out loud during the talk when I realized I was doing exactly what I would have told the class not to do—begrudge her for being someone who can help the team! I spent the rest of her time there welcoming her as an ally and someone who can help us fulfill on the bigger goal.

Don't be threatened by new hires who do what you do. It's actually much more useful to think of them as your allies. Befriend them, learn from them, and help them get up to speed quickly. They can help your projects get completed faster and better. They're not a threat.

It is with their help that you will have the bandwidth freed up to play an even bigger game and to take on what is next for you or to do your job even better.

If the company hires another thermal shielding expert, don't be threatened, be thrilled. Partner with them. Build them up; train them to do your work. They can take over your project and then you can look for what the next challenge is or you can work together to make your team the best in the company.

Measure your success not by how much a project needs you, but by how much you have built up the team such that it doesn't need you anymore. That is leadership. Then you are freed up to help take things to the next level.

You might also be jealous of people who have a nicer car, a romantic partner, a better title, or a cooler job description. That is normal. Just notice how it disrupts your appreciation of all the amazing things that you have and leaves you feeling insufficient.

The *Desiderata*, one of my favorite poems says, "If you compare yourself with others, you may become vain or bitter, for always there will be greater and lesser persons than yourself."

Celebrate your team's and even your industry's successes as your own. Our species benefits from all of these advances.

Jealousy is a wedge that keeps you away from people. Remember that a Jedi strives to remove any barriers between themself and another.

Coughing Up Furballs

My husband and I started a practice early in our relationship we called 'coughing up furballs.' Basically when one of us was upset, the other would sit down and ask them to cough up all their furballs, i.e. share what is bothering them. The calm person's job is to lovingly listen, without judgment or comment until they stop, and then to say, "What else?" (I prefer "What else?" to "Do you have anything else?" because the second is a yes/no question and feels like the questioner wants you to say, "No...") After they say, "what else?" they then continue to listen supportively without judgment or comment until they stop again. Then to ask, "What else?" until the person literally cannot think of another thing to add.

By that point the person usually is starting to feel better. They have been heard, listened to and supported and they are not walking around with an itchy furball in their throat! You don't have to have answers or offer your two cents, all you have to do is listen. You'd be amazed the difference it makes.

There are two essential ingredients to this exercise:

1. Finding someone you trust who can listen like that (or being someone who can listen to others like that).
2. Knowing they are not taking any of it personally or letting any of it impact them or their view of you. Their job is just to hold the trash can for your furballs and then throw them out.

If someone wants to vent about a co-worker or a meeting, let them. Give them a safe space to off-gas all of that negativity where it won't impact others. Be sure not to let it impact you too—remember you are helping them cough up furballs. Once they cough one up, throw it away. You don't need to hold onto any of it either!

[This is the exception to 'don't say anything you wouldn't want printed on the front page' rule. Sometimes you need to say things out loud to stop them knocking around inside your head.

Still be responsible with your words. There are ways to clear you mind without being needlessly harsh. You can say things that show you see your own role in things. Like: "I could see I was getting madder and madder at him because he wasn't listening to me the whole meeting and that irritated me because I hate not being heard!" versus saying, "He is such a ragging a$$#*^&, I feel like he needs a good kick in the teeth."]

Learning What Triggers Furballs

If someone is a chronic furballer you might say, "Sounds like that really got to you, do you know what it is about him or the way you two interact that triggers you so much?" They may just blame the other person, "Because he is such an idiot!" in which case, just leave it alone for now. But if they are willing to explore that with you, they might actually see that they hate being left out, not included in conversations, meeting or decisions. (Impressive if they can pinpoint it). Say, "Great; is that something specific to Andy or is that something that you had before this meeting?" If they can see that this is something that is not Andy's 'fault' but just part of their own makeup then they can start to learn to be responsible for it.

Same with you! When you get triggered by someone or something, go to someone you trust and talk it through, see if you can see what it is that bugs you. For me my triggers are very simple 1). Not being listened to and 2). People thinking I am dumb. It irks me to no end. "Brain size of a planet and you think I couldn't find the Yelp Review??? NO, I was saying I didn't see any information on the school there because you haven't uploaded any, not because I couldn't find the right URL!!!" But when I can know that that is just me and how I react and not them, then I can calm down, spare them the dramatics and say, "I'm sorry I wasn't clear, I meant that when I went to the Yelp page for the school there wasn't any information uploaded by the business and that we should update that."

When you know what your triggers are, you can be responsible for them. "Sorry that is just my judgmental/condescending thing. I apologize, that was not helpful. I am very grateful that you are here helping our team."

Being Responsible for Your Shortfalls

I think that to fulfill your mission, you have to embrace all aspects of yourself, especially your Dark Side. You have to be at peace and at home with who you are.

If someone says to you, "You are an impatient, self-absorbed jerk," you can say, "Yes, I am sorry I let that impact you. That is something I am working on. What can I do to make it up to you?" or, "Sorry, do you have any advice for what I could have done differently to not upset people?"

The common reaction to someone being critical or pointing out our flaws is to get defensive, offended or reactive. But that trains the people around you not to speak honestly to you, it puts distance in the relationship, it destroys trust and it leaves you and them in an unproductive funk after you walk away.

Just because getting defensive is the normal, average, or common reaction definitely does not mean it works! Aim higher. Embrace your whole self and you will be able to listen graciously to anything anyone says about you.

If you are scared to speak in public (like Richard Branson who used to throw up before going on stage), you can say before your presentation, "Sorry my hands are shaking, I get nervous talking in front of groups."

If you can own it, it won't own you.

Sometimes acknowledging the elephant in the room (that thing everyone knows but no one says) can help break the tension and put everyone more at ease.

When you can be at home in your own skin and be fully who you are in all environments and you grant others the respect and space to do the same, that is what I call success. You can be rich, powerful, famous, attractive, and successful but if you do not like who you are, you will not be happy.

Taboos

Many topics are taboo in our culture. We do not even dare speak about them. This is incredibly dangerous, as pain is increased by the isolation of hiding a secret. Many pains can also be healed by being spoken or shared in the right environment.

We often carry secrets with us of abuse, or mistakes we've made. We carry shame about our bodies or our sexuality. We hold fear of our inadequacies or being found out. We deal with the pain of loss or depression.

Some taboos are starting to be addressed in wider society, but within the space industry we are still lagging behind. The next sections discuss two taboos that are deadly to avoid talking about.

Suicide and Mental Health

"Then sometimes you'll play lonely games too,
games you can't win cause you play against you."
–Dr Seuss, Oh the Places You'll Go

Buzz Aldrin talks about his mother's suicide in his book *Magnificent Desolation.* He also talks about how he suffered from depression and alcoholism for years after he got back from the moon. I know two other NASA Astronauts from the Shuttle era whose mothers also committed suicide.

Suicide is a big deal and something rarely talked about. 800,000 people a year die around the world by suicide. **That is more per year than armed conflicts and natural disasters combined**. I don't want suicide to be a taboo. If someone is suffering, it needs to be ok to ask for help and we need to get better at providing help.

My friend and NASA Astronaut, Dr. Chuck Brady, who flew to space on STS-78 in 1996, died of apparently self-inflicted wounds in 2006 (five years after retiring from NASA). Yet, we still rarely discuss suicide and the pain and isolation involved and how we can reach out to help alleviate that stress and help people earlier.

Our culture of the old "Right Stuff" can make it even harder for anyone even to consider asking for help. People don't want to jeopardize their chances of being selected for a space mission.

I prefer to the premise that we all are struggling, all have challenges, all need support, rather than continuing a culture of, "I'm fine. If you need help, you're weak," which can feel like the unspoken culture of the west. What if it was standard operating procedure for all space staff and all space crewmembers to have a coach or a counselor to support them? We are involved in high-stakes, high-stress missions. We all need support.

I know many times I feel like I am slogging through mud, and that getting through the day can be an agonizing chore. It was more acute when I was in grad school and after the birth of my children, but it still pops up regularly, days where I just feel like I don't want to do anything.

We need to be talking about this reality instead of perpetuating the myth that we are all super-perfect. We need to give each other permission to be ourselves. We need to share our challenges and our struggles to help ourselves and also to help each other through.

Maybe my story can help you, or Buzz's sharing will make a difference for you. Our society is made stronger by sharing ourselves.

Living in space will be exceedingly challenging. The environment is completely unforgiving. We need to work out better ways to attend to everyone's mental health because one person can endanger everyone's lives quite easily in an off-world city. We can't afford to not get this right.

Even NASA's screenings and highly competitive selection process were not able to protect from mental health issues surfacing on the ground (one active astronaut was arrested in 2007 for attempted murder). Let's not count on "professionalism" to keep things working in space. This will not be sufficient to have teams working harmoniously over the long term. I have seen and heard of people losing it, even in very professional settings, far too often. We must aim higher.

We need to practice new levels of self-care, respect, communication, compassion and well-being: tools to make sure we are doing the best we can to support all people.

Homosexuality and Sexuality

The United States has gone from having sodomy laws on the books in all 50 states at the beginning of the space program to gay marriage being legal throughout the country (thanks to a Supreme Court decision in 2015).

There are openly gay men who have played in the NFL, fought in the U.S. Military and four who have openly served as Republicans in the U.S. House of Representatives, and still no space agency has ever flown an openly gay astronaut.

Sally Ride is the only astronaut I know to even come out as gay, even *after* their retirement from NASA. Even she only let this be disclosed after her death in 2012. I want to see our industry have more role models of people living 'out' lives, showing aspiring rocket scientists and astronauts that they can bring their whole selves to work.

One thing I am looking forward to is the number of out LGBTQ people who will get a chance to fly into space with Virgin Galactic, Blue Origin, SpaceX and the other space companies. We already have a number of LGBTQ customers willing to be ambassadors for space. I think this will be a great leap forward.

Sex in general is also a taboo.

I have talked to young women working at space companies who have had unwanted sexual contact and felt coerced and manipulated by managers and colleagues and then felt shut down for talking about it. But given how many careers have been destroyed, homes broken up and problems created from people having affairs, or harassing or assaulting others, it seems like something worth talking about, worth giving people the space to work through their concerns and challenges, and finding ways to mange their sexual lives in a manner that works.

We need to be able to talk about these issues openly if we are going to create a culture that works for all people.

Sex will happen in space and we need to be able to handle it responsibly and respectfully. It's part of our humanity.

Making Peace With Your Pieces

Your job this week is to face your Dark Side and know it is here to teach you something. When you are willing to go into your cave and face that instead of cower, the world will begin to open up to you. You will find a new access to making the difference you came to Earth to make.

Rather than something to shy away from, resist or suppress, turn towards those darkest aspects of yourself and shine the sunlight of love and compassion on them and see what happens.

If you want to reach your full potential you must make peace with your pieces.

"*The best way out is always through.*" —Robert Frost

Homework

1) **Practice noticing your Dark Side when it surfaces.** Catch yourself and take three deep breaths. Practice just letting it be. Not trying to fix it or change it or hide it, just sit with it or share it with someone else and see if it loses some of its power.

2) **Try to catch yourself throwing people under the bus.** Record it in your notebook.
3) **Interview someone in your life about you.** Ask them what you are good at and what you can work on. Take what they say so well that they feel safe enough to be more honest with you.
4) **Connect with a buddy and share what you are noticing**. Be sure to be gracious and see if you have more compassion for their Dark Side.
5) **Continue with your daily action.**

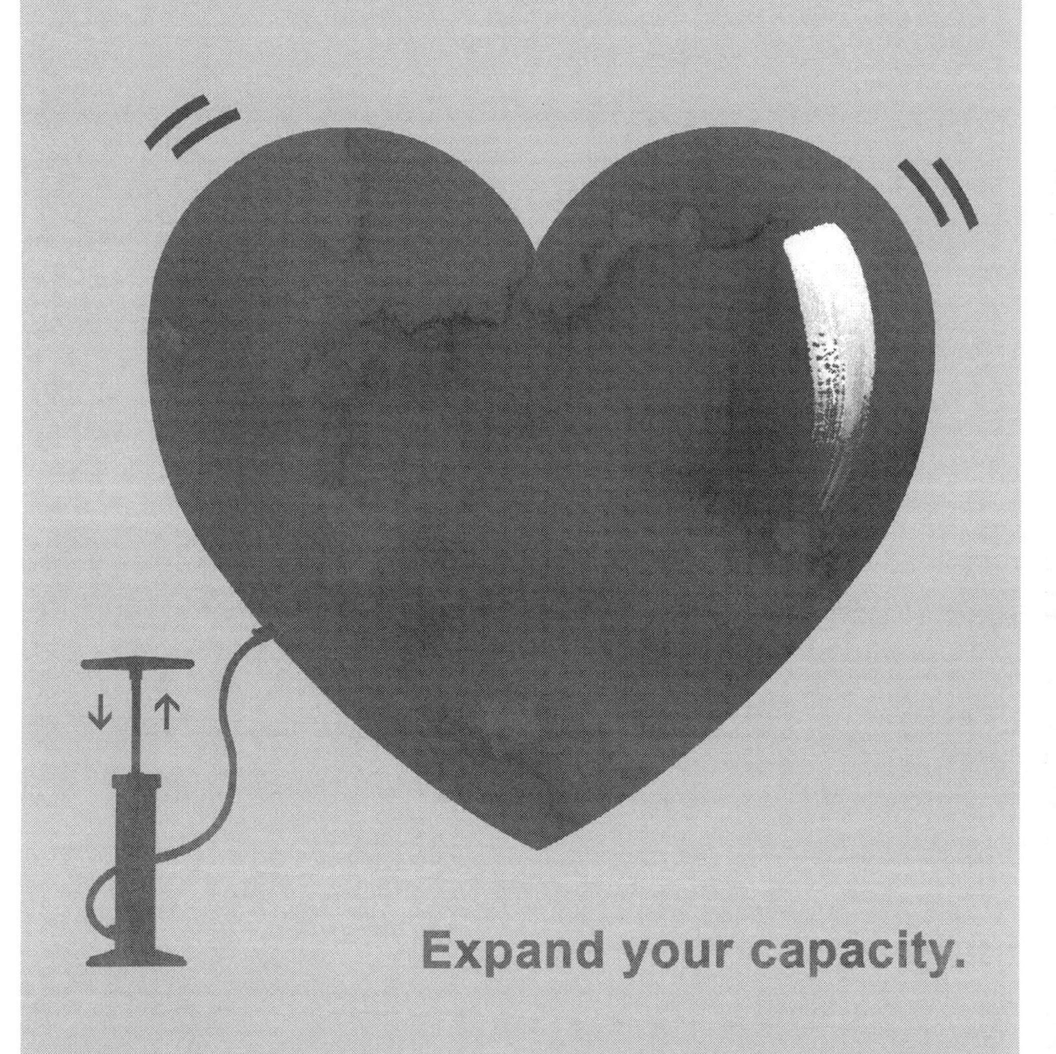
Expand your capacity.
JON MARRO

• CHAPTER 4 •

Do or Do Not, There Is No Try

You have answered the call to adventure, stepped into the Special World by committing to a Daily Action and faced your Dark Side. Now, like Luke Skywalker, you will go into the swamps of Dagobah to train for what is coming next, the Ordeal.

To do this requires a focus of mind. It requires a commitment to developing the inner strength and the depth of character that you will need to face whatever comes your way. This week we will practice these elements so that you will be grounded in your own power and ready to tackle the challenges ahead.

One place to start, if you want to train yourselves to live an extraordinary life of contribution and connection, is with the book *The Four Agreements* by Don Miguel Ruiz. Here is what the book says:

The Four Agreements

1. Be Impeccable With Your Word. This means doing what you say you are going to do, by when you say you are going to do it. It also includes not gossiping, or undermining others. This means not undermining yourself as well! Our words create our world. Words are magical and powerful, so use them for good.

In a space context, the First Agreement looks like delivering on your promises, even if it means working through the weekend or inspiring three other departments to expedite your project to make sure it gets done (and promising to bake them brownies for their efforts!).

2. Don't Take Things Personally. A lot of upset in life could be avoided if you don't take the things people say personally. Most times they are not responding to you anyway, but rather to some previous upset and they are taking it out on you. If you can keep that in mind you will not be the victim of needless suffering.

Not taking things personally looks like repeating to yourself, "This is not personal." And then asking yourself, "I wonder why they are hurting so much that they are even talking that way." And if you want to be a Jedi Master, "I wonder if I can help them get over that so they don't have to lash out at people that way."

3. Don't Make Assumptions. It may seem easier and safer (because you don't have to talk to anyone) to assume we know what someone means or what someone wants. But making assumptions can lead to needless suffering. It is much more powerful to be clear and direct in finding out what others are expecting from you and in making sure your expectations are clearly articulated for them. Clearly communicating can help you avoid misunderstandings, sadness, and drama. In the long run, that makes life MUCH easier *and* safer.

Practice direct communications. "I was disappointed that I didn't get to go to that meeting. What can I do next time to make sure I am included?" "You need this done by 3 pm? Is it ok, if I eat my lunch while I read it over?" "I am expecting all of you to be on time to the review, in your chairs and ready to go at precisely 3 pm."

4. Always Do Your Best. You may not keep all Four Agreements all of the time. But, if you do your best and don't beat yourself up when you fall short, you will find that your life will continue to open up and you will continue to grow without feeling self-critical, worried, or regretful.

"I completely forgot about our 11 am meeting. I got caught up in what I was doing and didn't have my phone with me so I missed the alarm. I apologize for leaving you hanging and I promise to check all my appointments at the beginning of the day and to keep my phone with me at all times. This will not happen again." Then make sure you do whatever is necessary to make sure it doesn't!

In the game of creating an extraordinary workforce, these Four Agreements are a great place to start. Imagine the difference we could make in the office, hangar, factory, lab, or classroom if we were all operating on this level and speaking openly about it. That would allow us to support each other in our efforts to master living in this way and working from these four agreements.

Do or Do Not, There Is No Try

"Words are, in my not-so-humble opinion, our most inexhaustible source of magic. Capable of both inflicting injury, and remedying it."

—Albus Dumbledor, Harry Potter and the Deathly Hallows

The first agreement is so important that I want to go deeper with it. Be impeccable with your word. This is critical to fulfilling your life mission. This is about integrity and learning to live your life by it.

I do not mean integrity as a moral thing. I am talking about the meaning of the word that is about wholeness and completeness. This is actually a physics conversation. Does this oxidizer tank have integrity? What will happen if it doesn't? Does the wing spar have integrity? What'll happen if it doesn't?

Take it to the next level: Does your relationship with your boss have integrity? Is there anything that needs to be said that you haven't said? Do you empower them? What about the other relationships in your life? Where is integrity lacking?

Without something being whole and complete it will not function properly. This includes test-fixtures, relationships and people.

If we want to make extraordinary results possible we need to demand from ourselves a higher degree of integrity.

When you say it, you are giving your word. All there is to do then is honor your word. "Do or Do Not, There is No Try."

Integrity also means telling the person you promised if you will not be able to meet the deadline or will not be on time. Give them as much notice as you can. Re-promise for when you can get it done or when you will be there.

In addition to the things you give your word to, there are also the things that people expect from you. You are accountable for seeking out others expectations of you and fulfilling them, unless you say otherwise. Why? Because when you don't meet other people's expectations it impacts their perception of you, whether they told you them or not. If you are committed to greatness, make sure you find out what people expect of you.

There is a lot that is normal in the modern world that we can "get away with" in this realm, whether it is being late, having a stain on your shirt, not calling your mom, not having a payment plan for your debts, or having cut off ties to a co-worker or cousin over a disagreement.

To raise the bar on yourself you have to clean up these areas of life, as a matter of integrity—of what works.

Cleaning up the things in your life that are out of integrity will give you the power and focus you need to face bigger challenges. It is great preparation for tackling your goals and your life. Bored of your problems? Get bigger ones.

Today we are going to write down everything that is lacking in integrity in your life. That light fixture in your house that needs to be replaced, the drawing that you promised you would finish

Monday, the "check engine" light that is still on, the person you still owe $60 to, that book you borrowed and never gave back. It's time to get all of it cleaned up. Start with making a list:

____

____

____

____

____

____

When My Integrity is Out, Drama Follows

In the Dark Side section I said that I can get the most angry/critical of others when *my* integrity is lacking. It is often something small that I am not really thinking about.

Like when my kids dragged Allen wrenches into my son's room and removed all the rungs of his bunk bed ladder. They were so proud of themselves! They told me they were engineers! I told them that engineers built things, not destroyed them and that if they really wanted to be engineers they would put the rungs back on the ladder. Then I left and went back to my computer.

A while later I heard a big crash.

I ran in and found that they had continued to take the bed apart and had now taken out the long screws that held up the cross bar of the upper bunk. Without the screw, the weight of the top mattress crashed down to the floor splintering the wooden cross bar under it. Now I was livid. I was screaming up and down, I was ranting, I was raving. “How could you do this? Do you know how much this bunk bed cost? I said engineers built things, not destroyed them!” On and on...

In the end though, my sanity prevailed. I realized I was most angry at myself. I knew they were doing something dangerous. I knew I should have stopped them, or at least supervised them and I chose to walk off. My integrity was out as a mother and there had been a consequence. Luckily for me neither of them were *under* the bunk when it had fallen. This was part of why I was so mad. If one of them had gotten hurt due to *my* negligence I would have been at fault and would have had to answer for that...

Now, was I *aware* of any of that when I was yelling? Not at all. In the heat of the moment, it was 100% their incompetence and insolence that was at fault, and I wanted to make it CLEAR to them, that *their* behavior was *not* OK.

A Jedi however must be more mindful. When you get angry, stop and look first to see where you are lacking integrity. How could something you did or didn’t do have spiraling into the upsetting situation? Where are you at fault too? It’s harder to let anger prevail when we realize that we are not blameless.

Next time you are really angry, stop and see if you can trace the upset to something you knew to do and that you didn't do (like stay and supervise the kids). Or trace the upset back to something you knew *not* to do, that you did (like leave a door propped open to a secure work area). If you can start to be honest with yourself about just that, and even admit that to the person you were angry with, you will stun and astound people. "I am sorry I am late. I shouldn't have stopped to send that one last email before leaving the office." Or "I acknowledge our team's test plan is late. I should have checked in on the status a week ahead of the due date and I didn't."

That is truly *not* a common thing to see in professional or personal interactions. "Sorry for getting so mad in that meeting. Jeff was right. I have been late on a bunch of drawings. It doesn't matter how buggy the software is, it's my responsibility to get them done and I didn't."

When was the last time you were angry? Where were you lacking integrity? (Where did you not do something you said you would do, knew to do, or were expected to do?) Try apologizing for that and see how life alters.

Leave It Cleaner Than You Found It

My husband George is the CEO of Virgin Galactic. When he goes into the bathroom at work and there is a paper towel on the floor, he picks it up. Why? Because he cares about the company and he wants everyone to walk into a clean bathroom.

When I see a paper towel on the floor of the bathroom at work, I sometimes have the thought, "Oh, that is the janitor's job." But the truth is their job is to clean the toilets and take out the trash. They have plenty to do without cleaning up the messes we leave too.

Use your paper towel to wipe down the counter. Pick up trash. Make sure you always take toilet paper from the smallest roll (especially one that is left outside the dispenser). If all four rolls are used equally, you don't free up a space for the janitorial team to install a fresh roll.

Start thinking on a higher level of what makes life work in the bathroom, in the break room, in the conference room and beyond. Operate like this planet is your spaceship and it's up to you to keep it working smoothly.

Power of Humor and Play

In our hangar it is important to keep people away from the vehicles who are not supposed to be near them and to keep FOD (Foreign Object Debris, aka trash, rhymes with 'odd') away from the ship that might damage it as it is taxiing out the hangar door or in flight.

We could have big signs with menacing letters reminding people to respect the stanchions that protect the ship. Or harp on them not to leave tools and small objects lying around. Instead someone, or a team, have over time added more and more funny signs around the parameter of the vehicles that use humor to make the point so much better.

There is one of Arnold Schwarzenegger dressed as the Terminator saying, "Who FOD'ed?" There is another that has Kermit the Frog dressed up in a smoking jacket sipping a glass of wine that says, "I saw you walk under the stanchions, but that's none of my business..."

People respond much better to humor and are much more likely to follow the rules when they are presented in a humorous and light-hearted way (think of Southwest Airlines' safety briefing).

Adding more play to the office lightens the mood, helps people connect on another level (like over the ping pong table), and helps them relieve stress. I am all for a few more humorous signs around the office (just don't make them at anyone's expense) and getting us to play games together. It's fun, and it makes us happier, healthier and more effective…

Using Inclusive Language

When you start operating at a higher level of integrity and thinking from what will make life work, you also start to be mindful of the words you choose. Words matter. A lot.

The current Oxford English dictionary notes that "man" and "mankind" are the old usages and that "human" and "humankind" are the new usages. Why are these new usages helpful? They allow us to be inclusive of all people. They make sure that everyone can be a part of the conversation.

Be sure to avoid gendered terms like "manned" or "unmanned" spaceflight. It is not too hard to use reasonable alternatives. You can say, "Our vehicle is piloted," that "it has a pilot and crew," that "we are about to start piloted test flights," and that "we are proud of our human spaceflight program."

When you talk about an upcoming job fair say that you need people to "staff" the table, or that we need people to "staff" the volunteer check-in tent.

I have heard people argue that the term "man" is inclusive and that it should be allowed. To make a point, a woman said that the word "woman" has "man" in it so it should be considered inclusive too. She said, "When we say 'women' just know that includes men too, ok?" to help men get what that logic feels like to others.

It's a funny thing; NASA has been using "Human Spaceflight" since at least 1996, when I first worked there. So it's strange working in the New Space industry, over 20 years later, and having people talk about "manned spaceflight" and "manned missions" at work or having journalists use those terms in stories in the media. I always feel like I am in the 1960's when I hear that.

Please do your best to find elegant, inclusive alternatives that honor our female pilots, astronauts and space travelers.

Privilege is Blind

I heard a man telling a story of being at a woman's conference. He said he was thinking about how he was forward thinking, inclusive, just-your-average-person, just-like-everybody-else, joe-average. He said one woman got up and said, "We all as women are united by our common struggle and our common goals. When we get up in the morning we all look in the mirror and see a woman looking back at us." An older black woman raised her hand and said, "when I look in the mirror I do not see a woman looking back at me- I see a Black Woman. And I have a whole world of that I deal with every day that you do not have to." The guy telling the story realized that *Privilege is Blind*. He did not see how much privilege he had as a white man, because he took his experience of life for granted and didn't realize that life is not like that for everyone.

So my invitation is to really be vigilant about being aware of how much privilege you have and doing whatever you can to reach out and make others feel included, welcome, appreciated. It makes a huge difference. Once trust is gone, it is really hard to get it back. Take care of people from the start.

Where can you see you might let privilege get between you and others? (Especially at work.)

It All Comes Down to Relationships

Relationships are the basis of any project, any team, any mission, and any company. If a relationship sours, it impacts the quality of the project and your quality of life.

Building strong relationships is as important as configuration management and documentation. It requires an investment of time, money, humility and directness.

There is a myth in business that if people are just professional, then things will work out.

I have not found this to be true. In every group I have ever been a part of, there are ebbs and flows of trust, cohesion, and comradery. The best teams are the ones that address any distance growing as early as possible and invest the time to work through the issue (often a misunderstanding) and rebuild trust and common ground.

Teams that try to operate on top of these rifts see tensions rise, discontentment grow, more mistakes, and more people suddenly finding "exciting new opportunities" in other companies. Many people say attrition is normal. I say it's a function of not investing in healing the disconnects that arise in work relationships. It becomes easier to change jobs and so people do.

It's a high bar, I know, and it does take a lot of time. But it makes for happier, more fulfilled people, and better products. It also opens up the possibility of a brighter future, a future where we can all thrive together—in the demanding confines of a space habitat or anywhere on planet Earth.

To me, that is worth investing in.

The Org Chart and the REAL Org Chart

In any organization you will likely find an org chat that shows how decisions, reporting and information is supposed to flow.

In reality though, as the dynamics shift between each of those people, the org chart shifts. People stop talking to each other. A team loses faith in their leader and start seeking guidance from another lead. What happens in the real world can be very different from what the chart says.

Some organizations will build an org chart to try to represent what is happening in reality. This at least is more honest. From there, you can see where the weaknesses in information flow are and work to get information flowing smoothly.

Projects rely on relationships and trust no matter what the org chart says. Learn to work out the real org structures and make sure the relationships in your organization are running well.

Answer the Question That is Asked

My husband will come home from work and say, “Did you schedule a doctor's appointment for Maya?” I should just answer the question that is asked with a simple, “No.”

Instead I’ll get all defensive, spun up and bent out of shape because of all the things I presume he is thinking. In my overactive imagination what he is really thinking is: “I ask you to do *one* thing in a day and you can’t even get that done?” What is worse is that my emotional reaction isn't to what he actually *said*, it is to what I turned it into in my mind.

So instead of simply answering the question that was asked, I am already halfway to a meltdown with justifications and defensiveness in my tone. "No, I haven't. Our son was screaming, my telecon went late, I was trying to get the Yuri's Night contract finalized and I am not even sure she needs to see a doctor. I'd rather wait another day and see if she gets better on her own."

A lot of escalations can be stopped by just answering the question that was asked.

In reality, he is not mad, he is not judging. He is just worried about making the appointment for our daughter. It would be more useful for me to say, "No, I will make a calendar reminder now to call when the office opens tomorrow. I am sorry I did not handle that as you were expecting." The anger, defensiveness and excuses will not help. Making a plan to handle the issue will.

This also works with questions you might perceive as dumb or simplistic. "Where should I put this completed request form?" Sometimes the snark really wants to jump out, "Oh, I don't know, how about in this basket marked, '*Completed Request Forms*'?"

No need to make people feel stupid; just smile and answer the questions that they asked. "It goes right here. Thank you for bringing it over!"

You will find your day will go ten times as smoothly, you won't get nearly as annoyed with how stupid people are and you will be happier from having helped so many people.

Remember: Answer the question that is asked.

Being on Time

I have never been a particularly punctual person. I like to time things to the minute to maximize the time I have at home before I have to walk out the front door. A Jedi however doesn't leave anything to chance.

What if there is an accident on the Metro or on the highway? Or I can't find my keys? Or I forgot my badge on the counter and have to go back for it? Any of these things can result in me being 5-15 minutes late. This may not seem like a big deal to you, especially if you work in a lax office place where meetings start at a leisurely seven minutes after the hour.

However, it is powerful to live as though it *does* matter. Punctuality creates a shift in people's trust in you. Walking into a meeting while someone is presenting is disrespectful and can throw them off. Think also of that person you know who is always on time, no matter what. You know you can count on them. "Randy? He will be there, he is always on time." Strive to be that person. The one everyone knows they can count on. That is Jedi.

When you arrive at the meeting earlier, you have time to do some pre-meeting conversations, which often have as much value as what you hear in the meeting. Plus you can be sure to get a good seat, have time to go to the restroom and grab some coffee before everything starts without being rushed.

Being on time in the morning is not just about getting up earlier. I think the most critical part of getting up on time is going to bed on time. I don't use an alarm to wake up, but I do use an alarm to remind me to go to bed! Punctuality (and power) start the night before. Create a powerful bed-time ritual and stick to it.

Once you have a great track record for punctuality, don't judge other latecomers. Remember you were like them until someone invited you to be your best self. Use your new commitment to being punctual to help raise the general level of performance of your whole team. It could even help you deliver your work products on time.

I think that is worth being on time for.

"Yes, and…" Speaking

"Yes, and…" is a powerful way to talk to people. It is a wildly underutilized alternative to "Yes, but…" or "No."

If the whole team wants to test a part only 12 times and you think it would be important to test it destructively as well, instead of saying, "No, I don't think 12 cycles is enough, we need to test it to destruction," say, "Great, yes, let's test it 12 times and then after we finish that I would like to run a test to destruction so I have that data as well."

Or at home you could say, "Yes, I would love to go visit your cousin Edna this weekend, honey. Can we be sure to be back by the time the game starts on Sunday and can we bring the mountain bikes so we can get in a quick ride while we are up there?"

Jedi deploy this kind of speaking mindfully. It is like the martial art Judo. You take the energy of your opponent and use it to get to where you want to go instead of opposing it.

Remembering People's Names

When I was working in the Astronaut Office at NASA Johnson Space Center fresh out of college, I worked briefly on a project with a few NASA Astronauts to evaluate medical hardware that was to be flown on the International Space Station.

Nearly six months later, I distinctly remember walking down the hallway in Building 4S (where the Astronaut Office is) and passing Mission Specialist Winston Scott who had been part of the medical hardware evaluation. As he passed me his face broke into a wide, warm smile. "Hello Loretta!" he said as he passed. I was 22 years old and far from anyone that he needed to remember, but he did and that has stuck with me all these years.

Many others I have talked to have also experienced Astronauts' uncanny ability to remember people's names. It makes people feel important, included, appreciated. It is something worth doing.

I was listening to the book *How to Win Friends and Influence People*. In it Carnegie tells the story of a man who never finished high school due to family circumstances. This man went on to great heights in his career because of his knack with people. He said he knew 50,000 people by name. If he can do 50,000 we can do 500!

Be the Astronaut/Jedi that you have always wanted to be. Devote more energy, time and attention to learning the names of the people around you. It will make a difference. You are not 'bad' at remembering people's names, you just haven't allocated enough bandwidth to the issue. Act like this is the person who will soon be interviewing you for your dream job, or someone you will ask out on a date. Give it that much attention.

First, ask people what their name is.

1. Repeat it back immediately "Nice to meet you, Brian."

2. Come up with a fun way to remember it like, "Monty Python's 'Life of Brian'!" (Linking it to people you love or knew as a child helps, too. "Just like that boy from 3rd grade Brian Carmody!")

3. Use it during that first interaction. "Brian, how long have you lived in Mojave?"

4. Use it at the end of the conversation, "Nice talking to you, Brian!"

5. Review all people you met at the end of the day or keep a list somewhere "Brian, structural engineer from Purdue, likes dogs."

6. Review it again before the next event. Test yourself to see if you can still remember all the names.

Remembering More Than Their Name

When you meet someone at a conference and they hand you a business card, write down what you want to remember about them on the card (or someplace you keep notes). Make sure you avoid writing anything that would throw them under the bus.

I met a friend Reehan who takes a selfie with everyone he meets. He doesn't use email, just Facebook to keep in touch since that usually includes your photo. I thought that was a very clever innovation as well, because recording the face that goes with the name is even more useful! Another way to do this is to make sure to get a group shot when a group goes out to dinner at a conference, or after an important meeting. Be sure to add a caption with the names of all the people in the photo. Not only are you recording the moment, you are also capturing the faces you want to keep with the names!

Help People Remember Your Name

Whether you are at work, at a conference or at a meeting, keep your nametag visible. It is best if it is above waist, pinned to your chest. If the font size your company badge is printed in is too small to read 5 feet away, print your name in a bigger font and slip it in the backside of your badge holder so people can at least read that when it flips to the backside.

Introduce yourself to new people and give them the mnemonic to help them remember your name. "Hello, I am Tony, like Tony the Tiger." When you see them again, you can say, "Tony Anderson, good to see you again!" In case they can't remember and don't want to have to ask you again.

As my friend Melissa Sampson likes to say, "Offering your name to someone, especially the next time you meet them, is a great gift and can remove the awkwardness from an interaction."

Help Others Remember Names

If you are hosting an event that has nametags on lanyards that may flip around, make sure to print two nametags and place them back to back in the sleeve or print them double-sided so either side shows the person's name. Use the biggest font possible for the first name; the last name can be smaller if space is an issue.

When you are with a group of people make sure to introduce them to each other, even if you think they've already have met. "Ryan, have you met Kim? She just joined the avionics team."

Learning people's names is the first step in creating community and team in your project, company or industry. It is a small step that can help us make a Giant Leap for humankind. Make it a part of your Jedi tool belt.

Graciously Accept Compliments

Our culture sometimes teaches us to be modest and shrug off or deflect compliments. "Really it was nothing," or "Zach did most of the hard work." Or "I didn't do all that much."

Try looking at it differently. Imagine someone is handing you a big present instead of paying you a compliment. All they want to do is give it to you. They're proud and excited to give it to you.

What would you say if your grandmother knit you a sweater that you didn't like and gave it to you as a gift? You would say, "Thank you very much Grandma, that is so thoughtful of you!" It doesn't matter what if you like it or not, that is what you would say. Give compliments the same courtesy.

It does not matter if you like the compliment or if you agree with it. Imagine it is a big present that they just handed you. Then graciously say, "Thank you so much, that is very kind of you to say." or "Thank you, I am glad you liked it." Remember the compliment is about *them* not you.

If you really had nothing to do with the report being done on time, you can say, "Thank you that is so nice of you to say. I will pass that on to Frank, Amanda and the team who pulled all the late nights this week getting this done!" Or if you had help you want to acknowledge you can say, "Thank you, I am glad you liked it. Abigail was a huge help on the project too. I could not have done this without her."

Generously Acknowledge People

It's nice to acknowledge people for a job well done. Sometimes a well-timed, authentic compliment can go a long way. Go beyond, "Good job, great work!" Try getting really specific:

"Thank you so much for getting that done for me by the end of the day. I don't know what you had to do to do it, but whatever it was I really appreciate it! You just made a big difference for me, for the company and even for my family who is going to get to see me an hour earlier tonight!"

Appreciating people for their work and their contribution can be as effective as a raise or a promotion (and you don't need management's approval to do it!). Give out compliments anytime you appreciate something. Make sure people know. It helps build your team and also helps positively reinforce the behaviors and attitudes you want to see around your project. It's a Win/Win!

You can even ask your colleagues, direct reports or romantic partners what they would like to be appreciated for. They might say, "I would like to be appreciated for doing the dishes last night." Or, "For not kicking Billy under the table when he started mouthing off in the design review."

Smile, nod, look them in the eye and say, "Thank you so much for doing the dishes last night and every other night that I haven't stopped to appreciate you for it. You are amazing." (Always go above and beyond what they asked for). Or "Thank you for showing incredible restraint and professionalism in the design review and refraining from physically assaulting our co-worker. Nice job."

Have fun with it. Make a production out of it. One of my favorite acknowledgements I have ever heard is, "And for everything you had to do to make that happen, that none of us will ever even know about: Thank You." It's a beautiful catch-all that acknowledges that they have likely done much more than they said. It's great to go ahead and acknowledged and appreciated them for that too. Go that extra mile. It makes all the difference to celebrate the good things people do when no one is watching. It affirms them and encourages them to keep it up.

There is a saying, "You get what you celebrate." Celebrate the little things and you will see bigger things to appreciate show up.

Ask To Be Acknowledged

Who doesn't like a good pat on the back? Who wouldn't mind a "thank you" for a job well done? Well, sometimes there isn't anyone around to see your amazing deed, or even if they saw it they may not realize what a big deal it is for you (or they may not be big compliment-givers).

That doesn't mean you can't still be acknowledged. All you have to do it ask for it. I don't mean asking, "Do you think I did a good job?" or "Do you like my outfit?" I mean saying, "Hey, can you tell me that I did a really good job getting all those reports to you on time?" or "Can you tell me that I look fabulous today?" Tell them exactly what you want to hear. (Think of it as thoughtfully making their job easier!)

The funny thing about compliments is that it actually doesn't matter that you asked for it explicitly. It will still give you a warm fuzzy to hear it. Imagine your romantic partner says, "Yes, you look great. That is a fabulous outfit and you look amazing as usual." Imagine your colleague says, "Thank you for working late every night this week so that you could get this report in on time. You are amazing." Imagine your friend says, "Thank you for agreeing to cat sit Amanda's cat while she is in Michigan so that I do not have to!" You will still appreciate hearing it, even if you told them the *exact* words to say.

So try it out. Try asking to be acknowledged. You can even return the favor by asking your friend what they would like to be complimented, appreciated, or acknowledged for and then giving them a fabulously generous compliment as well.

It's a great low-cost way to take care of yourself and everyone around you. (If this is way too far out of your comfort zone, start with asking fellow Jedi to give you a compliment. Another great reason to make sure you have people go on this journey with you!)

A New Hope

What if there were Jedi Knights at each space company and their job was to remind people of their greatness, remind people to put aside their differences for the bigger cause? What if we had a Pan-Industry Jedi Council? Or a Federation of Awesome?

That is my dream. It inspires me to imagine. I am committed to creating that kind of space community (and to you creating that kind of fun, high-functioning, transparent, supportive environment in all the communities that you care about).

It is already starting, and I get excited when I think about the sci-fi future that we will not only get to live in but also help create. All I have ever wanted—from the first time I saw *Star Wars*, probably before I could read—was to live in that future.

I can imagine a Pan-Industry Jedi Council convening regularly to make sure our companies are on track: finding the best synergies between our efforts, working through differences that arise, and training new Jedi in the ways of the Force. We will train them to speak powerfully and to honor what they say. We will train them to be able to see the best part of everyone and to speak to that part of each person. We will learn from them, too, because everyone has a lesson for us.

Homework:

1) **Complete at least two things on your integrity list.** Do the ones that will give you the most relief to have done.
2) **Acknowledge two people in your life.** Send them a paper note, an email or tell them what they did that you were impressed with, or why you appreciate them.
3) **Continue your daily action and buddy check-ins.** If you missed a day, just get back on the horse. No guilt, just resolve. Check in with your buddy too.

Bonus: Ask to be acknowledged for something that you did.

JON MARRO

• CHAPTER 5 •

Courageous Conversations

At this point in the Hero's Journey we must face death, or our deepest fear, and look it square in the face. It is in this crucible that the new leader is born. It is in the heat of the fire that a stronger you will be forged.

We each must confront a dark time, a time when felt like we could die. Those times when all hope seems lost, or when everything we knew how to do fails, a trial by fire. We each must face our own mortality.

You can do this by writing your obituary, participating in a high ropes course with a good facilitator, or by tackling a conversation with someone with whom you need to restore your relationship. Ideally, during this portion of your journey you will do all three.

Out of the moment of fear and even death comes a new life.

Writing Your Obituary

Start by getting out your journal and turning to a blank page. Write the date and "Obituary" at the top. Then write out an obituary as though you had died suddenly today—not like the New York Times would, but as it would be written in your own heart—with all your regrets and disappointments. Be gut wrenchingly honest with yourself.

Read it to a trusted friend to help you confront what you have just written.

Most of us live like we have an infinite amount of time to work on our relationships or to become the person we want to be. We are all also notoriously bad at tackling things that don't have a deadline (like these big life goals). Writing our obituary can help us face that our time here is finite and it's time to get moving.

It should sound something like this: "Gregory Lawton died today suddenly, leaving behind a wife who never got to spend enough time with him and two kids who stopped wanting to be seen in public with him about two years ago. He had a few projects out in the garage under a tarp that he never got around to finishing that his wife can finally throw out. He never visited Europe or got to see a Lakers playoff game. He probably could have stood to loose a few pounds. He is survived by his mother who he never quite forgave for making him wear hand me downs that were way too small for him and for leaving his father when he was only nine. He fell out of touch with his brother and they never made much of an effort to see each other. Their kids only met once.

"He was a manager of a space company with seven direct reports, but his career stalled out there and he was never sure why he kept being passed over for promotions. His biggest regrets were not flying his home-built plane to Oshkosh, Wisconsin for the biggest gathering of aircraft enthusiasts in the world and never fulfilling his dream of seeing the Earth from space."

Restoring a Relationship

When a relationship gets too tense, we need to do something about it. Sometimes that is shifting something inside our own thinking and sometimes that is speaking up and having a courageous conversation.

Before you give them a piece of your mind though, let's explore one of the tweaks you can try in your own mind to see if that helps.

Were My Insecurities Triggered?

When I was in junior high one boy taunted me nearly daily to tears for being flat chested. He would sneer, "You are flat as a board!"

Adults would say, "Ignore him, don't let him see that he is getting a rise out of you and he will stop." Their advice was even more frustrating! They were asking me to do something I had no capacity to do. No matter how hard I tried, I couldn't keep his attacks from getting to me.

Here is what I learned many years later: it is not what the bully is saying that is hurting me, it is what *I* am saying to *myself*.

His taunts hurt me so much was because they struck at the core of my insecurity—I wasn't as developed as the other girls. He was saying something I believed to be true and so I was hurting *myself* with that over and over. If he had been saying, "You suck at math. You can't even add two plus two," it would have been easy to dismiss the taunt itself because I knew I was good at math and had no issue with that. (Granted it never feels good to be taunted).

Also, if he had taunted me about math it would have been easier for me to see he was doing it out of his own insecurities and to have compassion for him. But because he hit on my insecurity about not being buxom I was too sensitive and too triggered to see clearly.

If someone is irritating you, see if there is something in what they are saying that is triggering your own insecurity. When you see it is your own beliefs about your flaws that are hurting you—not their words—that can free you up a lot. You have a lot more control over what you tell yourself than what they say. Work it out there.

There are times when we do need to address things with other people as well, so the next few sections offer insights into how to tackle courageous conversations.

When to Speak Up

"Tact is the knack of making a point without making an enemy."

—Isaac Newton

Deepak Chopra says, before you say something, whether in a meeting, at home, or in the middle of a courageous conversation, ask yourself three questions:

Is it True?
Is what I am about to say 100% true? Is he really *always* late or has he been late three times this week? Does the idea really "make *no* sense" or are there "parts of the idea" that I am "not comfortable with"? Is the person really "arrogant and disrespectful," or did they say something yesterday to which I took offence? Make sure you stop and check in with yourself about what is empirically "true" before you speak.

Is it Kind?
Some things are true, such as, "You are not performing up to par with any of the other engineers," or "The last three parts you submitted failed," or "Twice I've had vendors tell me you were rude." The next step is to ask yourself, "Is it kind?" Do you think sharing that will get the result you want?

Sometimes being kind can help you get a better result.

What if instead you said, "I have been impressed with how smart you are and how much you love space. I bet if you really put the effort in you could be one of the most respected engineers here in no time." Or, "I really admire your work ethic and your dedication. You've really streamlined a lot of our processes. I think if you can cultivate strong relationships with our vendors and earn their respect you will do well here." Try saying that and see what happens.

Is it Necessary?

Will it help the situation to say it or will it make things worse? This is an ego check. Sometimes my ego just wants to chime in, "Actually the first Space Shuttle flight was in 1981 *not* 1982." But in many situations it is not necessary to correct the speaker's error.

Keep your ego in check. Ask, "Is this necessary?" If you are training for the College Space Trivia Bowl the answer is "yes." If you are doing a design review on the injector valves, having the correct date is not necessary to the purpose of the meeting, so hold your tongue.

The Power of Being Kind

"It's nice to be important, but it's more important to be nice."

—John Templeton

Famed air show pilot Bob Hoover nearly died in a plane crash because a mechanic had put the wrong type of aviation fuel in his plane. When Hoover returned to the airfield, he walked over to the mechanic and instead of berating him, he said, "I want you to service my other plane tomorrow because I know you will never make that mistake again."

That show of kindness and goodwill helped turn a painful failure into an inspiring lesson learned. I bet that mechanic learned two very important things that day, what kind of fuel goes into what plane and the power of being kind.

The Power of Taking Responsibility

The next step toward speaking from a place of power and responsibility is to look for where you can be responsible for how things are going.

I was talking to a collaborator on the phone who was annoying the heck out of me. This is someone I work well with, have a good rapport with, someone I know my values are aligned with. Yet on that call I still ended up shouting at (however playfully and congenially) in my frustration. I even ended up hanging up the phone mid-conversation.

Instead of writing him off and encouraging others not to work with him or support his projects, I went to a neutral third party. I wanted help seeing where I was off-base, where I was being too reactive, and how I was being that made the conversation not work.

What came from that coaching conversation was the idea that I *hate it* when people don't listen to me. When my friend Metucka suggested that, I thought, "Yes, that resonates alright!" I can even hear myself in my head screaming, "You are not listening to me!!!" (Mostly at my kids, but I am sure I said it before they came along, too.)

I grounded in my passion for what is ultimately most important to me—the future of humankind. From there, I could give up my petty, egoistic, reactionary concern and I called my collaborator back. I apologized for letting my reaction be more important than our goals, reconnected at the level of our shared values, and worked together to find a solution for the project. I didn't talk about what he did, I talked about what I did and took responsibility for it.

So take responsibility for how things are going. Look for where you are off, where you have room to grow, or where you can be responsible for where things went off track.

This is especially important when it seems **obvious** that the issue is the *other* person's fault, when anyone around would agree with you that you are in the right and they are out of line. Then it is all the more important to see what you could have done better not to provoke them, to anticipate their upset, to respectfully address their concern.

They are not training to be awesome, YOU are. It's your job to make things work.

The Power of Forgiveness

There is a saying that resentment is like drinking poison and waiting for the other person to die.

Your grudge hurts you more than it hurts them. They may even be blissfully unaware that you are holding a grudge. You are the one that suffers. That is why forgiveness is not about them, forgiveness is about you. It's about setting yourself free. You do not have to agree with what they did, like what they did, approve of what they did, but I do recommend forgiving what they did.

Why?

Because it will set you free. And that matters.

Forgiveness is something you do inside you. You don't even need the other person there. Forgiveness is the process of letting go: letting go of hurt, letting go of resentment, letting go of judgment. They did what they did, you are where you are. It's all in the past.

By holding onto it, you can't move forward. You don't have to forget what happened, but you do have to let go of its grip on your life. When you do, you will feel like a new person. That is the power of forgiveness.

They already hurt you once, don't let the pain keep impacting you over and over again now. It's done. Let it go. No good can come of holding onto your pain, resentment and anger. The good news is that you have a say in that.

Your hurt thoughts and feelings are familiar, habitual, a part of your life, and it's time to let them go. Start by realizing that you are not your thoughts and your feelings, you are their observer.

Then start building a new life without hurt, a life you design.

The Power of 'I' Statements

The first step toward speaking from a place of power and responsibility is to speak from your own perspective: "I was hurt," "I was disappointed," "I felt passed over," "I was worried." Sharing your experience works better than being accusatory or critical. Avoid statements like: "You said that too harshly," "You give preferential treatment to the people in your cycling club and let them take the assignment I wanted," and "You were being reckless and irresponsible."

Even with an 'I Statement' you still need to be gracious. Saying "*I felt like* you were being callous and rude" is still likely to provoke a defensive reaction because you are being critical of them. Try, "I didn't like how I left our conversation earlier. I would love to get to know you better so I am less likely to get tripped up by misunderstandings. Want to grab a coffee sometime?"

You don't have to concede everything. You can also make requests. At the end of the conversation you can say, "Ok that sounds good. I can adjust to knowing that you raising your voice does not mean we're all getting fired. Can I also request you cut out the swearing? At least when I am around?"

Maybe just by you granting them space to be just the way they are (and not being critical of them) they might even surprise you and tone down their yelling too.

Conversation Pitfall: Triggering Defensiveness

The most common way to speak to people we disagree with is to imply that they are wrong. We do this so often that it seems like the normal, appropriate way to talk through an issue. "You shouldn't have been so unprofessional with me. I shouldn't have to hear secondhand that I have been taken off a project."

Even said nicely, those kinds of words almost always are met with a defensive reaction. You could hear back, "Well, I am sorry, but I was just tired of your attitude. It's been an issue on the project for awhile now."

Things usually devolve from there. Now both people are hurt and it only becomes harder to bring the two sides together.

When you are about to speak, make sure that you are not saying, implying, or offering anything that could even be misinterpreted as "You did something *wrong*."

What can you do instead? Look for where YOU can be responsible. This also gives you power. As Gandhi said, *"No one can hurt you without your permission."*

When you start to see yourself as the root of everything, you can see how you can turn around anything. "I am sorry that I did not make myself more approachable. What can I do better next time so that you feel 100% comfortable telling me bad news directly?"

If you turn the responsibility onto you, they have nothing to get defensive about. The issue is being addressed; you will likely get more candid feedback. They may even be inspired to take responsibility, too. They might say, "No, it's on me, too. I should have called you first. I was just taking the easy way out. I'm sorry."

How is that for a 180-degree turnaround?

Even if they don't step up to that level, they could say, "Yeah, you've been kind of frustrating lately, second-guessing all my calls. Maybe if you toned that down a little, we could get some more work done around here without the conflict." Mission still accomplished. The issue is discussed and you got honest insight into their experience of you. You even learned some things you can work on for next time.

Just remember not to get defensive about it. Your manager directly sharing their concern about you second-guessing them is a HUGE step forward. Be gracious to signal to them you can take them being honest with you.

Where Do I Start?

Before you do anything, you need to get yourself to a place of calm. It will be more difficult to bring your best Jedi to a conversation if you are upset, resentful, or vindictive. Start with coughing up furballs. Go to a trusted Jedi friend and ask them to let you clear your head.

Take a breath. Make sure you are doing this for the benefit of all humankind (the person you have the issue with included), not just you. Keeping the bigger perspective will help keep you from getting pulled into defensiveness and blame.

The next six steps are adapted from Judy Ringer's checklist for difficult conversations.

Step 1: Beginning the Conversation

You can try something like this:

"I'd like to talk about the team dynamic with you, but first I'd like to get your perspective."

"I need your help with what just happened. Do you have a few minutes to talk?"

"I need your help with something. I wanted to see if we can talk about it today/this week?" If the person says, "Sure, let me get back to you," follow up with them.

"I'd like to talk about our test plan. I think we may have different ideas about how to staff the test room, and I want to figure something out that works for everyone."

Write a possible opening for your conversation here:

____

____

Step 2: Get Interested In Them

Cultivate an attitude of discovery and curiosity. Pretend you don't know anything (you really don't), and try to learn as much as possible about the other person's point of view. Don't assume anything.

Work hard to give up your assumption that the other person is w-r-o-n-g and you are right. If you think they are wrong, it will come across in your tone, your body language and your demeanor and it leaves no room for magic. If you can be gracious and open-minded, anything is possible.

"I noticed that I wasn't included in the meeting invite for the future planning meeting. Can you let me know a little more about what I can do to get an invite next time?"

Pretend you're entertaining a visitor from another planet, and find out how things look on their planet, how events affected the other person, and what their values and priorities are there. Get interested in their perspective.

Watch their body language and feel where their energy is. What is important to them? What are they not saying? What are they worried about?

Sometimes I might even gently ask, "What are you not saying?" if it's a safe enough space for them to be honest (I usually do that more with people close to me who are just having trouble getting the words out. You don't want it to sound accusatory).

Let the other person talk until they are finished. Don't interrupt except to acknowledge you are hearing them. "Yes," "Understood," "I hear you," and "Thanks for letting me know," are good neutral ways of responding as they talk.

Whatever you hear, don't take it personally. It's not really about you. It is their story, their experience. Try to learn as much as you can at this point in the conversation. You'll get your turn, but be patient. Listen. See what you can learn. Listen the way you want to be listened to. If you listen with a closed heart and wait for your turn to talk, nothing will shift, nothing will alter.

What is this like for them? What pressures are they under? How is it for them dealing with you? Get interested in their side of the story. Let them say their piece.

Step 3: Acknowledge Their View

Acknowledge their view, show them that you've heard and understood what they are saying.

"Sounds like you have felt a lot of pressure to get the Master Schedule updated with every input and dependency and you've felt frustrated that the engineering staff keep changing their numbers."

Try to understand the other person so well you can make their argument for them. Then do it. Explain back to them what you think they said.

It doesn't even matter at this stage if what they said is not accurate. For example you can say, "So your experience is that I am late every single day. I can see how that would be frustrating!" even if you know for a fact that you were on time plenty of times.

The conversation is not about determining the facts. The conversation is about getting two people back on the same page. Facts do not do that alone "But I was on time the last four days- I can show you my clock in times!" does little to heal the mistrust that grew from someone perceiving you as always being late (we are not as rational and emotionless as we like to think. We are all subject to ego and emotion).

Once you hear them out, honor their perspective, apologize and get them to a point where they can even listen to you again, then you can show them the evidence you were on time recently. Until you get to that point though, those facts will not make a difference for them.

Acknowledging their perspective as valid can seem weird if we think it means we are agreeing with what they are saying. Keep them separate. You can honor someone's position without agreeing with it. Me saying, "you are frustrated that you have an employee who comes in late every day" or "This sounds really important to you," doesn't mean, "I agree with you," "You are right," or "I'm going to go along with your decision." It just means you hear them and are willing to listen.

This is an important part of the process. Acknowledge their view. Make sure they feel heard, respected and understood.

Step 4: Share Your Perspective

"You can't control other people's behavior, but you can control how you react to them." —Tamaira Ross, Blue Origin Engineer, Author of *Technical Leadership Begins with Knowing Yourself*

When you sense the person has expressed all their thoughts on the topic, it is your turn. From your perspective, what did they miss? Help clarify your position without minimizing theirs.

For example: "From what you've told me, I can see how you came to the conclusion that I'm not a team player. I think I am. When I raise concerns about the project, I'm thinking about its long-term success. I don't mean to be a critic, though I can see how I can sound like one. Maybe you can give me suggestions about how to bring up my concerns so that people won't take offense?"

Now is the time for you to gently, respectfully explain your position. Be sure to stop and listen respectfully if interrupted. Work hard to catch yourself whenever you get defensive (which is normally about every 15 seconds). Just keep calm. Let it go. (Or acknowledge it as discussed in the last section.) Keep the bigger goal foremost. This is setting a precedent for humanity that we can work things out, not about your ego.

A powerfully spoken apology is often a strong place to come from: "I am sorry I upset you and ruffled feathers," "I am sorry I judged you and wrote you off so quickly," "I am sorry I let things go that far, I should have spoken up sooner," or "I am sorry I didn't make you feel like you could come talk to me about it directly." Often taking responsibility for the breakdown can give the other person space to see where they might be willing to take responsibility too. But don't do it to get that result! Be willing to let them just accept your apology, too.

The beauty of it is that either way, you get to move from being a victim to a position of power. If you see that you had something to do with how things went, then you can be clear within yourself that this will not happen again. If you know yourself as someone who will speak up, who will work things out, who can basically handle any scenario, then you don't have to worry about what anyone does to you. You are in charge of your life.

Step 5: Brainstorm Solutions

Brainstorming and continued inquiry are useful here. Ask your partner, "What do you think would help us?" or "What ideas do you have that could make this work?" Whatever they say, find something you like and build on it.

If all of their ideas are actions for your team with nothing that their team will do different, say, "Great, I think we can find ways to implement a lot of those ideas. If we did, could we also count on your team to do X and Y?"

If their idea doesn't solve the problem, skip being critical, just add what will help. Try something like, "We would be happy to loop in the requirements manager at the beginning of the process, I still have a concern though about being able to keep this on budget, what ideas do you have for that?" or "Can you also get me weekly budget updates to make sure that is on track?"

If the conversation becomes adversarial, gently ask them, "What is most important to you about this project?" or "What do you most want to make sure gets done?" or "How did you first get interested in space?" (i.e. get interested in them and their point of view, find common ground).

Asking for the other person's point of view usually creates safety and encourages him to engage. If you've been successful in staying calm and centered, adjusting your attitude, and engaging with inquiry, building sustainable solutions should come relatively easily.

The hardest part is usually just getting to the table and starting. Once you've both had a chance to feel fully heard and their experience was validated, it is usually straightforward to come up with ideas for what you can do together going forward.

Step 6: Practice, Practice, Practice

The art of conversation is like any art–with continued practice you will acquire skill and ease.

Here are some additional tips and suggestions:

A successful outcome will depend on two things: how you are being and what you say. How you are being (centered, curious, supportive, problem-solving) will greatly influence what you say.

Acknowledge emotional reactions–yours and your partner's–without judgment or blame.

Know and return to your purpose at difficult moments.

Don't take verbal attacks personally. Help the person you are working with come back to the goal of creating a workable relationship.

Don't assume your partner can see things from your point of view.

Practice the conversation with a friend before holding the real one. Ask for feedback on how to be more generous and gracious.

Mentally practice the conversation. See various possibilities and visualize yourself handling them with ease. Envision the outcome you are hoping for.

What If They Really ARE Wrong?

"Those who cause suffering are suffering." –Buddhist saying

If you are having trouble finding compassion for someone you want to talk to, because they "really *are* a jerk," remind yourself that hurt people hurt people. If they tried to hurt you or someone you love, it is probably because someone hurt them.

Try walking in their shoes. They are tired, misunderstood, alone, worried, frustrated. Maybe they are even dealing with something much worse that you don't even know about. An aging parent, a troubled kid, a medical diagnosis, a messy divorce, their own troubled past.

Give up that you KNOW who they are and why they are doing what they are doing. Instead, get interested in who they really are and how you can help them. (If they stop suffering, they might stop causing suffering for others.)

Great, now you are ready to go back to Step 1 again. Get Interested in Them.

Homework:

1). Write your Obituary. Share it with a trusted friend or your buddy. Think about how you would want to live if this was your last year on Earth.

2). Have a Courageous Conversation. Pick a relationship you've been avoiding, or trying to ignore. Do it. Create a result you would be proud that you accomplished.

3). Continue your daily action and check in with your buddy. Support them in taking on their courageous conversation and doing their daily action too.

JON MARRO

CHAPTER 6

Jedi Ways of Being

Living in World 2

When I was in college, my best friend and I had a concept we called World 2.

It basically described that place where time stops, where all your real-world (World 1) concerns disappear and all that's left is what really matters. Often we would be in World 2 during late night conversations in the dorms sharing our hopes, our dreams and our fears. World 2 was the place you could open up and be yourself. Where you were loved and accepted just the way you were. It was a place I first found at summer camp.

It is a magical and intoxicating place.

During a high ropes course I was reminded of my goal in college of getting to LIVE in World 2 all the time. I also had a goal of helping others to experience its magic as well. On the ropes course, we worked hard to keep ourselves present and stay in conversations about how we wanted to grow as people, about our challenges and goals and about what it would take to get there.

I was so happy to be there that my cheeks hurt from smiling so big. It really is my happy place, people connecting, opening up, being honest and courageous. I invite you to bring a little more World 2 to your life—to the people closest to you and even to your co-workers.

Instead of small talk, ask people:

-What is one thing you love about your life?

-What are you looking forward to?

-What is something you are proud of?

Conversation is an art. So let's paint! What are you creating? If Gene Roddenberry can create an inspiring vision for the future of humanity, we can, too. Let's build a World 2 we all can live in, one conversation at a time.

Finding Your Niche

In every company, project, or planet you have a unique role to play. Your job is to figure out what it is and fill it.

Part of that is listening. Listen to what is wanted and needed around you. Listen to your heart. Listen to what the universe is telling you. This is why journaling, meditating, and walking, hiking, or running can be so useful. To be able to hear, you have to give the muse quiet times to speak to you.

I have heard it said that praying is talking to God and meditation is listening.

Think also about what you love to do. One thing I love to do is connect two people who I think need to talk. If someone says, "I am really passionate about real estate and am super excited to learn more about it!" I would tell them, "Oh! You need to talk to Rafael; he lives near you and knows tons about the local market and what you would need to know to get started! Plus you both love soccer, too." To me making a good match is a puzzle to solve, a game, something I love to do just for the sport of it.

After I read Malcolm Gladwell's book, *The Tipping Point*, I learned there is even a name for what I do. He calls it being a Connector. (In the book he also describes Mavens and Persuaders.) Being a Connector is something I can do for my community or company to be a 'value added' contribution. It's part of my niche.

What is your niche? What is your unique spin on your role? Do you love to present? Get a team of people all to support the same solution? Do you love sifting through giant data sets? Do you love working with your hands? What are the skills that come naturally to you or that you want to develop that augment your academic training or your job title?

You might love to play guitar, speak to kids about space, help meetings move quickly and efficiently, catch errors deep in the code, streamline processes, or help departments play better together. Whatever your unique skillset or passion, own it, develop it and find opportunities to use your superpower for good. Use it to help as many people as possible. Is playing guitar at happy hour part of your job description? Is streamlining Human Resource's onboarding paperwork part of your role as an avionics engineer? Usually not, but don't let that stop you. The people who make the biggest difference in the world are those who are willing to give their talents wherever they are needed.

Volunteer, step-up, run something, suggest something and offer to start it up yourself. Make it so. Don't wait for someone else to ask you. They don't know all your awesome hidden talents, passions and skills—you do.

Running a Lean Meeting

After I left NASA's Johnson Space Center, I had the privilege of working for Dr. Chris McKay at the NASA Ames Research Center in Silicon Valley. Chris is an astrogeophysicist who has spend a lot of time in the Antarctic Dry Valleys and is one of the world's experts on terraforming Mars. He is also the most Jedi boss I have ever had.

Chris is a huge champion for just about everyone and does a great job of bringing people to science who would not otherwise get a chance. He is also very good at listening. He looks a little like Daniel Day Lewis in *Lincoln* and acts a bit like that as well.

My favorite thing though was to watch him lead a meeting.

He doesn't believe there was any reason to have a meeting longer than it needs to be. He says it comes from a deep respect for people's time. If someone in the meeting was talking too long without moving the ball forward he would skillfully, artfully, interrupt and ask what the action item needed to be and who would be taking it on.

It was fun to watch him work. Just at that moment when everyone in the room *knew* we were about to waste time, he would intervene and save us all and use our time well again. Somehow that gave us a great feeling of being honored and taken care of.

It is such a fantastically useful skill that I wanted to share it here to invite you to practice noticing when a meeting is going off the rails and thinking of what you could say to re-focus the conversation on what it needs to accomplish.

Chris said the key to being able to skillfully interject is to fully listen. If you are Jedi Listening to what they are saying as well as the energy behind what they are saying, they will feel fully heard. Then you can interject because you fully listened to their communication and they can leave feeling heard rather than shut down. This can be hard because when they start "going on and on." You want to stop listening, when instead you really need to tune in, embrace it and then lovingly interject.

Chris said to listen to them like they are the son or daughter of someone you respect very much. And to listen to them not from, "what value are they to me," or "how can they help forward my goals and dreams," but to listen to them as valuable in and of themselves — to listen to all beings respectfully and fully.

Just talking to him, you can feel the difference it makes.

Then you could say, "Thank you Colin for doing all the thinking on that. Would you be willing to write up all the concerns you mentioned into a memo and sending it around to all the team?" or "Great Liam, so it sounds like we need to order the controller from all three suppliers and test which one will work best in our experimental set-up. Can you make sure that gets done today?"

Of course his other advice was just to not have meetings in the first place. Chance encounters in the hall or walking over to someone's desk was often a more efficient way to following up on an action or a project or get buy in from everyone on the team on a new idea.

Happy People Do Better Work

I also laughed as I confessed to Chris McKay that it took me the full year of working for him to realize that when he said he would be at the "Hydrolab" from 1 pm-2pm, that he wasn't referring to some mysterious lab he had in another building, but rather that he was at the lap pool NASA inherited on the Navy side of the airfield.

He was unapologetic. He said, "Happy people do better work. People need to take care of themselves, whether that is time with family or friends, running or swimming laps."

He makes a good point. It's the, "Put on your own oxygen mask before helping others," rule from flying on airplanes. If you are taking care of yourself, eating well, getting outside, sleeping, and maintaining meaningful relationships you will be in a better position to contribute at work and to the rest of humanity as well.

It's your job to attend to your own well being first, and then help your team and your family to make sure they are as well. Because happy people do better work.

Be Willing to Say "I Don't Know"

How many times have you been in a meeting and someone uses an acronym or a term that you don't know? Often times we make a note to look it up later so we don't have to stop the meeting and we don't have to admit we don't know something.

There are two problems with this:

- It often makes it harder to understand the rest of the presentation without knowing that term.
- There may be others in the room who also don't know what it means and are also now not following the conversation.

If you think you are in either of those two situations, be courageous. Respectfully ask the speaker to clarify what they mean.

If it's a high level meeting and not critical you are following everything in the moment and it would be counter productive to stop everyone to get your question answered, you can also wait to ask a trusted colleague at the break.

As a corollary, when you are speaking, avoid acronyms and terms people in the room might not understand, and when you are a program manager, avoid inventing new acronyms. Especially avoid inventing acronyms just to fit a word you like (unless it is really funny, like The Late Shows' Stephen Colbert, who made up the Combined Operational Load Bearing External Resistance Treadmill (COLBERT) to name the new Treadmill being launched to the International Space Station)

Similarly, if someone asks you to code something in Python, don't say yes and then order *Python for Dummies*. Be clear that you haven't programmed in Python before but that you are excited to put in the extra hours to learn it (if you are). Say that it might take you a little extra time but that it will be worth it to them to have you trained up on it and ready to go.

It may feel awkward to admit you don't know something. Maybe even something you should know. Keep in mind that however awkward that is, it is better than feeling like you are hiding something and like you are a fraud. Instead just own what you know and what you don't know and be clear you are committed to learning and improving all the time. If you are honest, confident and hard working, that will go a long way.

Receptivity: Tuning in to the Universe

One of the metaphors I appreciated a lot was that we are antennae picking up on what the people around us are feeling and on what the universe is sending us. (I like to picture a little radio dish moving around to pick up a signal.)

Next time you are in a weird mood, or feel 'off,' consider that it may not be personal. It may just be something you are picking up in the airwaves. Maybe it is not *you* that are depressed or stressed — maybe it is someone around you.

I used to get mad at myself for feeling that slogging-through-the-mud feeling. I thought it meant I was a failure or not built as well as the clear thinking people who could just work hard everyday.

Now I realize it can be many things, a reminder from the universe to slow down and take care of myself. (Even computers need to be re-booted every so often to keep working properly!) It can also be a Bat Signal from someone in need. Maybe the way I pick up the signal is to feel terrible myself. Maybe that is a way to get my attention to go help someone else.

Actually, the ONE thing I have found that reliably makes me feel better on days like that IS to talk to someone more upset than I am. By listening to them and coaching them my over-concern about myself disappears; I appreciate getting to contribute to someone else; I take my own coaching and I feel better.

This is a powerful Jedi Mind Trick. Next time you feel bad, consider that it is not personal and that it may be others' moods you are picking up on.

Thinking Original Thoughts

Make one of your goals to think a new thought, something that has never been thought before. It will add a new level of "awakeness" to your speaking and help ensure you do not get stuck in ruts. Your brain is just like your quadriceps, you have to use it or lose it.

Physicist David Bohm, whose doctoral work was taken from him and used to make the first hydrogen bomb, in later life wanted to come up with a way to heal the rift that was growing up between people. He promoted something called a Bohm Dialogue, which is where a group of people sit together and think. No one says anything until someone thinks something that they have never thought, read or heard before. It is such an intriguing concept I wanted to pass it on.

Next time you have an original thought, celebrate it. Train your brain that that is something you value and appreciate and more will come your way!

Appreciating Where You Get to Be

Sometimes when you get up for work, it can seem rote. There is nothing new or shiny or exciting. It's just the same thing you did yesterday.

In 2002, we had music superstar Moby come to NASA Johnson Space Center in Houston, Texas. He spoke to over 700 young employees in Teague Auditorium (at the time I didn't even know we *had* that many young employees around NASA Johnson!).

He said to them, "Driving into the Center this morning, I was so inspired and excited to see a real Saturn V rocket on display at the front gate and to know that this is the place where human spaceflight happens. Now I know that for you driving in that gate every morning, it doesn't feel like that anymore. But that doesn't mean what you do here isn't amazing. It just means that you have gotten numb to it. That is ok. I am numb to the amazingness of my life too. To me, being a rock star is like walking into a different stadium full of lights and crowds every night and thinking 'oh another stadium.' But I just want to remind you that this is not normal, this is awesome."

It was an extraordinary moment and a great reminder that what we do everyday may start to feel routine to us, but it never stops being extraordinary.

Live Like You Are On A Spaceship (You Are)

When you live on a spaceship you take special care of where every last one of your fingernail clippings goes. You never leave a mess in the bathroom.

You treat the place as if you are the one responsible for it and no one is coming behind you to clean up (or if they are, you don't want them to find your fingernail clippings that have floated into the air intake filters…)

Now try treating your home planet with the same respect and see what happens. Leave places cleaner than you found them. Pick up litter on your hike. Throw out the paper plates after the company break room party. Treat Earth like it's your spaceship—because it is.

Leave restrooms cleaner than you found them. Why? Because it's your spaceship. (You are just about to wash your hands anyway.) Just shift the way you relate to an overflowing trash can in a public restroom from annoyance to an opportunity to contribute. (Just use your clean paper towel to push the rest down and no one else has to get annoyed that the trash can is overflowing. You will be a micro hero!)

Use less water, even when there isn't a drought. On the space station the crew can't shower the whole time they are up there. Maybe we can get by with a little less water down here as well. Be mindful of water. Treat it like the precious resource that it really is. On Mars you won't take it for granted (at least not for the first 100 years ;)).

It can be a lot of fun to be the commander of a spaceship. Might as well start practicing now…

Be Yourself, No Matter What They Say

My friend Stephanie Abler in high school had a harsh way of reminding me not to worry about what others thought of me. She used to say:

"You wouldn't be so concerned with what people thought of you if you knew how seldom they did."

I was pretty self-absorbed at that age (possibly still…) and she was trying to get me to see that most people didn't think about me nearly as much as I did. She was actually a genius—other people are much more worried about what everyone else is thinking of *them* to think much about me.

It's a nice reminder to "be yourself, no matter what they say." (Those are the lyrics to a Police song that I like to sing when I doubt myself.)

So relax. Don't worry what everyone else thinks. If you are being true to yourself and your integrity is in, you have done the hard part. The rest will work itself out.

Being Charismatic

Charisma is being powerful and fully present — fully in the moment. You don't have a secret agenda, and you are not just biding your time until you can move onto something more interesting. It is the way of being of someone who has arrived, who doesn't need to prove anything, someone who is completely at home in their own skin (usually faults and all). You are not trying to be cool, or impressive. You are just being there 100% with the people around you, giving them your full attention.

Yoda is a good example of this, so is the Dalai Lama. Astronaut Shannon Lucid would also be another great example…

For me, being completely at home in my own skin took awhile. I had to deal with worrying that people would find out that I am a fraud (classic impostor syndrome). Then I had to deal with worrying about them finding out that I am attracted to women (it might be hard to imagine being stopped by that in today's world, but remember I am a product of the 80's). Now it's starting to get better, but I still have to deal with self-doubt and awkwardness around some things (like trying to explain that I am a self-proclaimed Jedi Trainer for Space Heroes.)

Giving up that I don't have somewhere more interesting to be is also hard. I have a ridiculous ego and I rush to judge everyone and everything around me. So if I am at a reception I practice focusing on the person I am talking to and give up that I would rather go over to the hors d'oeuvres table again. I also don't look up to see who is walking by (usually!) and just give who I am talking to my full attention. (A great practice for developing charisma.)

Being Audacious

When I was about to finish college, I was scared out of my mind about what I was going to do next. I had no ideas. I had a degree in biology but no thought about how I could use it to get a toehold in the industry of my dreams, the space industry.

I remember sitting down in the lounge late one night in Terra (the co-op I lived in) with a wizened grad student and him saying to me, "You are about to graduate from Stanford! You can do *anything* you want!"

I hadn't really thought of it that way. I had been thinking more along the lines of, "What could I possibly do?" and here he was reminding me, "You are awesome, you are in demand, you have a lot to offer!"

Now you don't have to be graduating from Stanford to think like that. It's just a way of reminding you to be audacious, to think big, to think out of the box.

Then he said, "So what do you most want to do?"

I thought about it and I said, "I want to work in the Astronaut Office in Houston." "Great, call them in the morning." So I did. The woman on the phone had her answer pat: "I am sorry ma'am, NASA is in a hiring freeze right now and we are not accepting any applications."

I said, "I know, I want to come and work for free."

She did not have a pat answer for that. In fact, she paused a moment, unsure what to say. Luckily for me, she passed me on to Mike Kinkaid in the Co-op Office who listened to my story and decided to take a chance on me. I started at NASA Johnson Space Center (JSC) as an unpaid intern in the Astronaut Office in the fall of 1996.

Thanks to Mike, I was adopted by the NASA Co-ops (college students who alternate working a semester at NASA and a semester at school). I had the time of my life — largely because I finally found a group of friends that loved space as much as I did!

I rented a room in someone's house for cheap and got rides to and from work every day. I lived on the cheapest food you can imagine and ate over at friends' places at lot. It wasn't easy, but it was a lot of fun.

I went to lunch with them every day, got to tag along on their special tours and lectures (like one from Gene Krantz on failure not being an option) and plugged into their social network.

We had a ball. The co-ops I met at NASA are still some of my favorite people in the world. We all had dreams of making a huge difference at NASA and we supported each other in our audacious aspirations. Because we ate lunch together every day and worked in departments all across the center, we often had our finger on the pulse of what was going on around JSC more than our management did.

When the center director said he wanted to have the whole center stop work for a day to host industry leaders and showcase our technology to them, the idea didn't get a lot of traction. Most NASA staff had never heard of Industry Day, and if they had, they didn't care much. At a Space Shuttle crew return party we set out to offer NASA Johnson Space Center Director George Abbey help getting the word out.

Now to a 22-year-old, George Abbey can be a pretty imposing figure. He is a heavy-set man who never appears to smile and he was notorious for tightly controlling who gets to fly into space and who doesn't. As future astronaut hopefuls, none of us were eager to get on his bad side.

I tried asking him to dance but he declined. My friend Leeward was more successful at engaging him. He told Abbey that the Co-ops felt that the center was not getting behind Industry Day and that we thought we could help. The very stoic Center Director swiftly invited Leeward up to his 9th floor office in Building 1—the unreachable upper echelons of power to us lowly Co-ops— and asked him to explain.

In the end, Abbey agreed to our plan and at 6 a.m. one morning, dressed all in black, an army of Co-ops gathered, divided up the map of the sprawling campus, and set out with armfuls of flyers and tape. Our mission? To post a newsletter letting people know about the importance of Industry Day on the inside of every bathroom stall door at JSC, the one place we knew people would read it.

We were bold and audacious and we loved every minute of it.

Being Fun

Back in 1997, in addition to Industry Day, NASA Johnson Space Center had another day that they asked everyone to stop working for every year. It was called "Total Health and Safety Day." I am sure whoever started it had very good intentions, but as a fresh-out I was very frustrated that an entire day would be devoted to safety and our focus wouldn't be on how to avoid another *Challenger* disaster, or to "never forget," or that space is hard and we need to make sure we were vigilant in always seeking out every possible failure mode.

No, "Total Health and Safety Day" seemed to me to be mostly about how to reduce tripping hazards around the office, lower your cholesterol, and wear seatbelts. All good causes to be sure, but not something I felt was best served by a whole day of employees wandering by booths set up around the campus.

I decided to take matters into my own hands and volunteered to run a “Morale Booth” at the fair. I felt that morale was one of the biggest Health and Safety issues at the center and that if we could keep our morale high, we would do a better job of communicating with each other and avoid the kinds of culture breakdowns that can lead to the endangerment of our space crews.

More urgently morale was also the biggest risk factor for the young set I was working with. We were not at risk for heart disease and were not driving forklifts or lifting heavy boxes, but we were idealists coming to NASA with a dream in our heart, and I saw time and time again that flame get snuffed out within weeks of arriving at NASA. I felt that it was important to our “Total Health and Safety” that we help each other make sure that flame never went out.

Our Morale Booth had candy, NERF machine guns, "Dilbert Bosses" to use for target practice, handouts with an inspirational foreword from Gene Roddenberry (the creator of *Star Trek*), and a tent full of young inspired Co-ops happy to remind you why you wanted to work at NASA in the first place.

It was really one of my favorite contributions to NASA and I hope it will serve as a reminder to you to bring fun to your workplace!

Being Creative

Currently there is a movement in education to push STEM subjects: Science, Technology, Engineering and Math. There is a group that has been working to put an "A" in STEM to make it STEAM. They want to add Arts to the movement.

Naturally we in STEM fields felt a little concerned about this. Wasn't the point to focus on math and science? Wouldn't adding 'Art' dilute that?

Then I heard a talk by a woman from the Los Angeles County Museum of Art (LACMA). She asked us, "Do you want to train a generation of scientists and engineers, or do you want to train a generation of Leonardo da Vinci's?"

That got me thinking. I do want a generation of Leonardo da Vinci's. I also want to encourage the engineers and scientists already working with us to develop their creative side. To let loose. To make art. To make music, to dance. To think out of the box, to be creative, to Think Different. That is where the magic is, that is where the breakthroughs are going to happen.

My co-op friends and I had a mentor at NASA Johnson, Dr. Kenneth J. Cox. He was the Chief Engineer for the whole center. He even had a fancy office on the 7th floor of Building 1. He was a delightful older man, who while very slender, had a Santa Claus-like twinkle, a short white beard and a great belly laugh.

We somehow managed to follow him to some Creativity Conferences in Houston and just had a delightful time mixing their world with our dreams of spaceflight. Often the tools of social scientists, storytellers and creatives are dismissed by engineers or scientists as woo-woo or not mission critical. Yet, as we deal with more and more complex problems and are confounded by the inevitable human element that comes up in teams, you will find the things you learn hanging around creative types may just be what gets us to Mars.

Your Attitude Determines Your Altitude

One of my previous jobs was as a Flight Attendant for Zero Gravity Corporation. I would be on board the plane supporting our customers as they got to experience weightlessness for about 30 seconds as the plane flew a 10,000 ft roller coaster in the sky. Since the plane does 12 Zer0-G Parabolas, we get to float around in the plane for about 6 minutes per flight.

One day I showed up for work not too excited about the day ahead. I realized that was unacceptable given the incredibly cool nature of my job—getting to float around in a 727! So I got together with my colleague and fellow Jedi, Tim Bailey, and we decided this was going to be the Best Flight Day Ever!

We started going around to all the other staff, pilots, mechanics, coaches and letting them in on our plan. "Hey everyone, today is going to be the Best Flight Day EVER!!!" As we went around from person to person we got more excited and so did they. We started to create a buzz out of nothing and soon everyone was having fun and trying to think of ways to make this the Best Flight Day Ever!

By the time the flyers arrived we were dancing and hooting and hollering and they were happy to join in the festivities. There is even a great photo of me, that ran the next day in the Washington Post, welcoming flyers at the base of the 727 aft air stairs, hands in the air, head thrown back having the time of my life.

I went on to do a few dance routines for the entertainment of the flyers (and the staff) and we went on to have a really fun, amazing, low-stress flight day that truly was my best flight day ever.

That was not how I showed up to work that morning. It was a completely created reality. That is the power you have in any moment, to create and to be fully alive— just because you can.

Homework:

1). Practice gratitude. Every time you have a complaint, stop and think of three things you are grateful for. Write them down.

2). Write a Thank You Card or Note to someone who made a difference in your life. Be as generous as possible in your letter. Cite specific things you remember them saying or doing that are important to you. Send it without expecting a reply.

3). Continue your daily action and check in with your Buddy. Support them in their daily action. Share your gratitude lists.

PLENTY
GRATE FULL
FULLFILLED
JON MARRO

CHAPTER 7

What You Came To Earth to Do

"At the climax, the hero is severely tested once more on the threshold of home. He or she is purified by the last sacrifice, another moment of death and rebirth, but on a higher and more complete level." Based on the Hero's Journey, by Joseph Campbell.

In *Return of the Jedi* Luke has a final dual with Darth Vader on the Death Star:

"I know there is still good in you," says Luke as they dual. At the end of the scene Vader ends up turning on his master, the Emperor, and saving Luke's life. As Vader lay dying in Lukes's arms, he tells his son, "Tell your sister you were right about me. You were right…"

Most likely your dad is not quite as aligned with the Dark Side as Darth Vader, but most of us are still harboring resentment, frustration, or judgment about one or both of our parents. For me the message of Return of the Jedi is, "Everyone is redeemable."

To complete your Jedi training you must take on bringing love and honor to your relationship with your parents. If Luke can do it you can too. Hopefully you won't have to wait till your dad is dying in your arms to get there. [Kylo Ren side note: If you don't forgive your parents you are more at risk of turning to the Dark Side!]

Can you see that there is still some work to do in creating the relationship that you want with you parents? Are you willing to do that work over the next two weeks?

Redeem Your Parents

When I was in my 20's I was not willing to work on my relationship with my dad at all. I was more than happy to wait until he was on his deathbed to talk through anything with him. Luckily for me at 27 my Jedi training intervened. I thought my dad was an idiot, an embarrassment and not good enough for me. I also thought he had had an affair when I was 12 and so had been giving him the cold shoulder for 15 years. I didn't like being around him, wanted nothing to do with him and had no interest in doing anything to change that.

Jedi training taught me that I couldn't complete my own training without restoring my relationships with my parents. So at 27 I called my dad and had that conversation that I hadn't planned to have with him until he was 80.

It was one of the hardest things I have ever done in my life. The old-school landline handset felt like it weighed 100 pounds.

I apologized for being such a jerk to him for the past 15 years. I told him it was because I had thought he was having an affair and that I wanted to have him back in my life again. He said, “Kids come up with the damnedest things,” "The door has always been open," and that all he wanted was to, “have his little girl back.” That Christmas when I was back home we went out for dinner at Applebees, just the two of us, and talked for three hours.

Turns out he wasn’t so bad after all…

My life has been forever altered from that conversation. My kids got an awesome grandpa and I got the possibility of a romantic relationship where I wouldn’t kick the guy out eventually for not being good enough!

Whatever your beef is with mom or dad, get over it. You turned out great. The only thing they had to do was give you life, anything beyond that is gravy. So lighten up on them. I promise you it is hurting you a lot more than it is hurting them to carry around hurt, resentment and judgment. It leaves distance and it cuts you off from your full power. Start experiencing all their quirks as their way of saying they love you.

If anyone at work does that thing your dad or mom does, it irritates you and gets in the way of your relationship with that person at work.

It can also get in the way of romantic relationships. I had a string of relationships that would start off great and then I would decide they weren't good enough and walk. It wasn't until I got reconnected with my dad that I saw I had thought, "He wasn't good enough." After that I was free to make a relationship actually stick, so I asked George on a date.

If you think your parents are great, then look at anywhere where there might be distance in the relationship, anything you haven't told them, anything you would regret not having said or asked while they were alive and have that conversation. You can always call and thank them profusely for all that they did for you.

Your relationship with you parents is the foundation of all other relationships in your life. At work, romantically, with your own kids etc. The better you can make it, the better all your other relationships will be.

You assignment in this section starts with taking on your relationship with your parents… It is a pre-requisite to fulfilling your full potential in this world and certainly for fulfilling your mission.

It's Time

For about a year, I have been hearing in my head the phase, "It's time," over and over again. It's a powerful call to action. You've trained, you've worked hard, now it is time to take action, to go out into the world and fulfill on what you came to Earth to do.

What Is Your Mission?

In Section 1 you had an opportunity to create a draft of your mission. Mine was "Using space to bring out the best in humanity." Now you get to dive a little deeper.

It's time to look at your lifelong mission as the context for your whole life. When I think about my life in terms of what I am here to do, it helps take me from the hour/day/week perspective of life (which includes unreturned emails, laundry, and making a dentist appointment) to the year/decade/lifetime perspective (which is looking at my goals, values and priorities).

What happens when I look from there? My priorities may shift; another value or goal may stand out more. I may need to change some of my behaviors (like undermining my kids), or add in some new habits (like reaching out to someone new everyday).

It is also from the big perspective that I draw the energy to give someone a second chance, or to listen to them/respond to them more generously or even to work harder, get up earlier, make that phone call I am avoiding. I may not "want to" in the moment, but if I ask myself, "Is doing it consistent with the world I am committed to?" If yes, then I better give up that I don't want to!

So what is your mission? What does the world need that you came here to provide?

If your mission is to go into space, or run your own company, take a deeper cut. Why do you want to do that? This is even bigger than your "Dream Big" dream, this is why that dream matters to you in the first place. Why is getting the first picture back from Alpha Centauri important to you? Why is making access to space affordable important to you? What does your Big Dream enable for humanity? Keep looking until you get down to the fundamental concern that is driving you.

Often we settle for a reason we have heard, "for science," "for survival of the species," etc. These certainly could be your reasons, but make sure they really work for you. Make sure it's a good fit. If we brought no research equipment, would you still want to go to Mars?

Ask yourself the tough questions: Why is humanity's survival important? What about us is most worthwhile? It is time to really do the thinking. If humanity turns out to become a solely destructive force in the universe, is your top value still its preservation? Try to articulate what it is you want to save about humanity. If your goal is space access, what do you want to enable for humanity and why? If we open up space as a terrifying new theater of war, would that be the fulfillment of your vision? Get really clear on what you want.

If you want to be an astronaut, why? Beyond being your dream, what does its fulfillment enable? Coming back and helping people all around the world have the courage and the determination to make the difference they came to make, too? Coming back to Earth and being an advocate for clean water, or human rights? Coming back home and giving back? Think it through. What difference do you want your flight to make?

If you want to run a space company, why? Could you make your space company a model for how to bring out the best in people while taking care of the home galaxy? Could you make a space company that makes a lasting difference in people's lives? Or that helps companies better work together for better efficiencies in a hostile environment?

Many of us have a drive to be important. That is ok. But this is a bigger game. How can you use that drive to make a difference?

Think of it as not just *what* you are going to do, but also *why*? Splitting the atom is interesting, but also dangerous. Do you want to do it just because you are curious or because it will be a great accomplishment? Or do you think that what you are doing will fundamentally help make the universe safe, happy, just, and inspired? What if you turn out to be wrong?

Think through the why. Why send humans to Mars? Why develop new rockets? Why send probes to the stars? If you don't think deeply about why you want to do it, it is more easily misused. If you are going to space for peace, or to help humanity learn to get along, do you accept military contracts? What about military contracts from powers hostile to your government? Where do you draw the line?

My goal is to get you thinking beyond yourself, beyond achievement, recognition, winning, accomplishment, and wealth. I have never seen any of those lead to true happiness on their own. Only when they are accompanied by a commitment to contribution and self-improvement will you really experience joy.

This game is much bigger than just you. It's a game that is as big as the universe itself. So take two, what is your mission?

____

____

____

____

____

If you really like what you come up with, consider adding it to your email signature. It will remind you of who you are and it might just inspire some others too!

Finding Your Purpose

Your purpose builds on your mission. It talks about how you will feel when you achieve your mission and by extension how you wish all people will feel.

My mission again is "To use space to bring out the best in humanity." My purpose is that "All people experience being fulfilled and connected."

You can see that my purpose doesn't even have space anywhere in it! At first this was hard for me to accept. I am so all about space. But really at the end of the day, even for me, there is more to life than just space and what drives me at my core is people experiencing being connected and fulfilled.

No matter what projects I am involved in or what I do, the question for me is always: “Will this help people feel more connected at the heart-level, more known for who they *really* are?” and “Will this help people fulfill what they came to Earth to do? Will this give them a deep sense of fulfillment?”

It is why I am writing this book. I want YOU to be fulfilled!

One way to get at your purpose is to ask yourself, what would I feel like when I have accomplished my mission? (When I am walking on Mars, when we send off a starship to Alpha Centauri.) Answers could be: fulfilled, connected, happy, self-expressed, inspired, present, grateful, loved, peaceful.

Then ask yourself: is that something I would be happy just having for myself or is that something I would like the whole world to experience? Would you bc satisfied if you were the only one who was happy, fulfilled, connected, self-expressed, inspired, etc.?

Consider that your real purpose is that after you fulfill your mission, not only do you feel happy, connected and self-expressed but that you continue to work so that ultimately everyone else in the world does, too.

Exercise: My purpose is that all people experience/are:

____

____

____

Do the Most Important Things First

A teacher came in to teach about time management. She brought out a gallon jar and put four big rocks into it. Then she asked the participants, "Is it full?" They said yes and then she got a bag full of pebbles and poured then into all the spaces between the big rocks. "Is it full now?" "Probably not…" said the students, catching on. Then she brought out a bag of sand and poured it into the jar and it filled in all the spaces between the pebbles. The jar was now filled to the brim with sand. Then she got a pitcher of water and poured it in.

"What is the lesson?" She asked. "You can always fit more into your schedule!" a student answered. "No. That if you don't put the big rocks in first, they will never fit in."

In the story the big rocks are your priorities, the most important things to you (spending time with family and friends, developing yourself as a leader, working on something that really matters to you). If it is important, do it first. Do it everyday. Make your day consistent with your priorities.

In the end your life is just a collection of days. The laundry and the dentist can all fit into the spaces in between.

Put the big rocks in first.

Ask for What You Really Want

When we were at the United Nations UNISPACE III conference in 1999, the Space Generation delegates were asked to submit a list of recommendations to the U.N. of what we wanted them to do.

I wanted the world to have a holiday where we, as a planet, celebrated the power of space to bring humanity together. I proposed it and it quickly got adopted into the list of ten proposals the Space Generation Forum would present to the U.N.

I was thrilled.

The U.N. liked the proposal so much they decided it should be a whole week and that the week should go from Oct 4-Oct 10 — the anniversary of the launch of the Russian Sputnik satellite to the anniversary of the signing of the Outer Space Treaty. They named it World Space Week and the world has been celebrating it ever since.

The problem was that that wasn't what I had meant. What I *really* wanted was for the U.N. to declare April 12th a global space holiday. April 12th is Cosmonautics Day in Russia. It celebrates the day the first human, Yuri Gagarin, flew to space: 12 April 1961. April 12th is also the day that the first Space Shuttle launched, in 1981. I was getting ready to launch Yuri's Night a new global holiday that celebrates the power of space to bring the world together and wanted the U.N.'s backing.

The only problem was, I hadn't *mentioned* to them that I wanted the global space holiday to be April 12 — and so it wasn't.

I learned a valuable lesson that day that I have never forgotten. Don't be afraid to stand up and ask for what you really want. Be clear, don't mince words, and make sure you are understood.

Note: 12 years later, on the 50th anniversary of Gagarin's flight, the U.N. did declare April 12th International Day of Human Space Flight.

"So, What Do You Do?"

I was at a leadership class in L.A. and I met an amazing, powerful, and charismatic man. I couldn't help but ask him what he did for a living. He answered proudly, "I am accountable for the safe transport of 40,000 kids to their classrooms every day within the LA Unified School District." I was impressed, thinking to myself, "Wow, the district is lucky to have this guy." Then he smiled and added, "I am a bus mechanic."

He taught me a lot about how you carry yourself and how you hold what you do in life. Whatever your job title or role, be proud of it. Be clear about how it maps to what you are up to in the world. If you are driving for Uber, say, "I run a small transportation company by day, and am also working on writing my first book."

If you don't like your title, make up a new one. "I am the Creativity Lead for our Rocket Propulsion Test team, my job is to come up with as many ways as possible for our rocket to fail on the test stand so that it doesn't fail on the launch pad."

Remember you are not your title or your job. You are who you say you are. You can say, "I am accountable for making sure the pointy end of the rocket always stays pointing forward!" or "I make sure that all of the instruments on board the ship are always talking to all of our computers on the ground." Since every single system on our vehicle (and some off) is mission critical, you don't have to exaggerate to talk about how important what you do is.

"I am accountable for testing every type of material that we put on our vehicle so that we know its limits and are flying within them." Have some fun with it.

Get Bigger Problems

One of my favorite pieces of advice is that if you want to get rid of your problems, get bigger problems.

If you are overwhelmed because you have to do laundry, you have to go the store to get groceries, you haven't even started on the analysis that is due Friday and you really want to go for a run, try this. Sign up to run a company wide event, sign up to speak at a conference, agree to raise money for a kid with cancer, or volunteer to mentor a FIRST Robotics team.

You can't get rid of problems. You can however exchange in problems you have had for awhile, problems that are boring you, no longer helping you grow, and exchange them in for bright, shiny bigger problems that will really test your mettle.

You might also find that when you need to fly across the country for the FIRST Robotics national finals, that laundry stops being a problem, it's just something you handle to make sure you are there and available to help those students.

Imagination is More Important Than Knowledge

I was at the California Science Center for their SpaceFest and the Chief Scientist from NASA Armstrong was giving a presentation.

On his first slide was the famous Albert Einstein quote, "Imagination is more important than knowledge." What struck me, though, was that he had posted the next part of the quote as well, which I had never seen before:

"Imagination is more important than knowledge.
For knowledge is limited to all we now know and understand,
while imagination embraces the entire world,
and all there ever will be to know and understand."

–Albert Einstein

Let that sink in. Remember to always keep your mind open to all that still lies beyond what we now know.

I think if we just told kids all the things adults DON'T yet understand (and that is a long list), they would be way more interested in learning than just talking about the great many things we have figured out.

The Future Doesn't Just Happen, It Has to Be Created

When I was a kid, the year 2000 was a long way off and very magical. We were certain that by *then* we would have flying cars and a base on Mars. When it didn't happen, I realized that the future doesn't just happen, it has to be fought for and created every step of the way.

Look around the room. Every single object around you was once just an idea, a concept in someone's mind that they had to make real and convince others was a good idea. They had to "make it so."

So think about the future you want for yourselves, your children and grandchildren, for humanity and think about what we need to be doing now to get there. Let's do the thinking for the future we want, design it and fight to make it so…

The Real Reasons We Go Into Space

Former NASA Administrator Mike Griffin gave a speech in 2007 about the 'Acceptable Reasons' to go to space and the '*Real* Reasons.'

"In my view, the space business more than most other endeavors suffers from the fact that the most important, the best, and the most basic reasons for doing it are Real Reasons and not Acceptable Reasons. The Acceptable Reasons — economic benefit, scientific discovery, and national security —are, in fact, completely correct.

But they comprise a derived rationale, and are not the truly compelling reasons. And again, who talks like that, about anything that really matters to them?"

"All of us here tonight got where we are by being analytical and objective and very left-brain oriented. Spaceflight can't be successfully accomplished without these traits. And so I think we tend not to pay appropriate respect to the deeper parts of human nature, which are intuitive and qualitative."

He goes on to talk about the Real Reasons:
-Wanting to strive to go beyond ourselves, to push the boundary.

-He talks about curiosity and the power of the awe and wonder of the cosmos and of seeing something no one has ever seen before.

-He talks about building a monument — like the cathedrals of Europe. How that required people to make technical innovations in building tall, strong walls and roofs that can span large distances that enable the western world to develop. It also , required innovations in *ideas* that moved humanity forward— like people delaying gratification by taking on projects that span generations and creating guilds to teach trades.

Then he ends with this powerful thought:
"When you do things for Real Reasons instead of Acceptable Reasons, you have a chance to obtain Real Success."

So let's practice articulating the Real Reasons we want to go to space at every opportunity! Learn to get good at it, explore more powerful ways of expressing your Real Reasons for going and then have the audacity to say, "Yes, that is why we are doing this."

Going to Space is Like Having a Baby

Having a baby doesn't make a lot of economic sense. In the U.S. it costs about $250,000 to raise a child. They are not going to help you with $250,000 worth of chores on the farm and most likely won't take care of you in old age either (previous practical reasons to have children).

Yet we still insist on having children. Why?

Because there is something bigger at play than raw economic benefit. We are investing in a future; we are doing something we believe in, something we derive pleasure and satisfaction from. Maybe we too are learning to delay gratification and think about projects that take generations to complete.

On a pure biological level, you can say that we are fulfilling a primordial drive to reproduce, to propagate our genes and our species.

Space is just like that. We go, even if the economics don't make sense on their own. We are compelled — to bring about a future we believe in, to seek to fulfill a grander mission, as well as to propagate our species and ensure its survival.

We Need to Learn to Be Better Parents

Last time humanity explored new places and built colonies in remote outposts it didn't go so well. Colonies eventually had to fight for the right to govern themselves.

This time, when we create outposts in distant places we need to think of them as children, not colonies. They do not belong to us; our job is to help them grow up to be self-sufficient and productive members of society and then to let them go.

As I was yelling at my son, however, I realized that perhaps my "parenting model" for space settlement is going to depend on us learning to become better parents first. Can we learn to empower and love unconditionally? Can we learn to have them turn out without hating us?

Parenting has never been my strong suit. I was intrigued that it was now tied to our future success in space. If we want to master the human interactions necessary to have our space settlements prosper and thrive without wanting to cut off ties to, or undermine the home planet, we have to master the human interactions in parenting.

There are a million modes of human interactions we will need to master to thrive off-world (where any one person can easily sabotage the whole system) and parenting is very much one of them.

So parents, the challenges you face everyday raising your offspring, are also great training for your space future. Take heart, it is for a good cause you are doing this! And for goodness sakes, please get all the help you can to learn the tools of Positive Parenting or anything else you can get your hands on to empower you to be the kind of parent any kid would be proud to have.

Let's work to be awesome parents to the 1st generation of space settlements. It's the best gift we can give the future.

Vulnerability Is The New "Right Stuff"

American author and social critic Norman Mailer wrote a book called "Of a Fire on the Moon" in 1970 about the Apollo moon landings that predicts the loss of interest by the public because of the "remorseless banality of the astronauts, and the fearsomely conformist culture of NASA itself."

A new thought struck me while reading that. Maybe the courage and bravery that will be required of our next generation of space explorers, scientists and engineers is not just the willingness to risk your life, the real courage required is the willingness to be vulnerable, to share ourselves fully and to let the world go on the emotional journey with us, too, not just the physical one.

Maybe if we are honest, open and fueled by a love of humanity, we could better connect with the hearts and minds of the public. If we are willing to share our true selves despite all the pressure of our culture to be rational or reserved we will make a more profound difference for our species.

Ideally, we will even make the very difference that we all hope space exploration is capable of creating. Namely, the building of a mature, connected, responsible species that any of us would be proud to send to the nearest star system.

My goal is that we build the culture of the Sci-Fi future to go along with the cool spaceships. To create a real Starfleet or Jedi culture we will need to work on opening up more with one another, sharing our true selves, not just our work selves.

That is why vulnerability is the new Right Stuff. It is what it is going to take for the next generation of space explorers to get us not just to the stars, but also to an inspiring future for humanity.

We all have a lot of training to do. This is going to be even harder (and more rewarding) than you ever thought.

In addition to being able to handle being strapped into a small spaceship atop a large explosive device you also need to be able to share about the experience meaningfully and inspire audiences with what they are capable of, too.

Challenge Accepted.

Building Your Own Jedi Circle

It's great to read inspiring books, to watch insightful videos and to work by yourself, but there is also a real power to working in a group.

To get to the next level of your training and development, you need to be a part of a community. There is something that you can get from working in a group that you can't get any other way. If you have benefited from this journey, consider starting a group at your work or in your community, or even a virtual group to support each other in taking on what is next.

It will take putting yourself out there, making invitations, risking rejection and sharing your vision, but those are all great opportunities as well. It doesn't take much. Meet once a week, once a month or once a quarter, or even once a year. Create goals, intentions, outcomes, for your group and a format. You can have people share, you can discuss a book, you can organize exercises, or even outings or field trips to do random acts of kindness like brightening strangers days with street corner dance parties, or chatting up strangers and believing in them. You can rotate who leads the conversation, keep it casual, or invite in guest speakers. Whatever you do will be the perfect thing. It can also change over time to meet the group's evolving needs.

Create a group that celebrates each other's victories, and supports each other through setbacks and failures. These people may become your closest relationships in the world.

The Rite of Passage of Initiation, like completing your hero's journey, has you welcomed back into a community that honors, recognizes and understands that you have changed completely. Create a group that will celebrate who you are becoming. It will serve you and your mission to build a circle around you like that.

Using the Force of Knowledge and Defense, Never for Attack

"A Jedi uses the Force for knowledge and defense, never for attack." —Yoda

A Jedi is very perceptive of what people need and how to talk to them to create a particular result. That is a very powerful skill. It could easily be used to manipulate people or take advantage of them.

Just like in Harry Potter's "Defense Against the Dark Arts" courses, powerful magic can be used for the Dark Side. It is your job as a Jedi to only train people who are committed to using their powers for good and to insure we ourselves do not slip into the temptation to use the tools and training we have been gifted to serve our own egoistic drives.

Passing It On

"*You teach best, what you most need to learn.*"

—Richard Bach, *Illusions*

The best way to learn is to teach. As you are developing yourself in your new Jedi skills, like Jedi listening, not throwing people under the bus and finding the good in anyone, it works best to pass it on.

When you find someone who is interested in growing and developing themselves, take the time to get to know them. Earn their trust. Ask if they are interested in your perspective. If they are receptive, use questions to draw their mind to what you are thinking.

"Andrew, what was another way you could have asked for the day off?" Get them to figure it out on their own. Anything they come to on their own will be ten times more powerful than anything you or I tell them.

Remember they have just as much to teach you as you have to teach them, so stay humble, and stay open to their coaching and their contributions to you. Share yourself, your journey, and your growth with them. Give them challenges to work on and always acknowledge and celebrate their successes and their efforts.

Follow Your Dreams and You Will Give Others the Courage to Follow Theirs

Many of us space people have longed to be astronauts, to float around in the peaceful vastness of space, to bound gracefully across the plains of the lunar surface, to climb the mountains on Mars. But don't let the intensity of the longing fool you.

"Getting there" does not bring happiness.

Just ask Buzz Aldrin. When he got back from the moon he had to confront, "What do I do now?" "How do I top that?" For over a decade he struggled with depression and alcoholism. Just "getting there" was not enough.

That doesn't mean that the longing is misplaced. It is a powerful force that you can use to make amazing things happen in your field and in your life. It is also a gift to be celebrated. Having drive and direction is wonderful; just ask someone without it.

So there are two things to remember:

1. *Enjoy the ride*. Enjoy getting there. Enjoy the adventure, the people you meet, the things you do to prepare yourself, everything you can create to make your dream real. Make it as much of a joy as the actual flight.

2. *Put it in perspective*. Put your goal inside a bigger goal, like "help the world feel more related" or "open up space access to more people." Make your bigger goal one that will take you longer to achieve, so when you get back from the moon you are clear about what there is to do next.

Consider that the Universe gifted you with this longing, this drive, in order to fuel your bigger contribution to humanity.

Now you have the adventure of creating what that contribution will be. Consider that it is the contribution that you leave to the world that in the end will be the true source of your happiness, and not the time you spend floating weightless. The beauty of your gift of ambition is not just that it will get you to Mars, but that it will also open up this larger opportunity to give back to humanity. Think about what you want it to be.

This larger opportunity ties back to your purpose, too. Your bigger mission could be to come home and help others find that, too. That is really 'paying it forward.'

Just remember that your time in space marks not an end in itself but rather the beginning of an even bigger adventure.

Follow Your Dreams and You Will Give Others the Courage to Follow Theirs…

Homework:

1). Make peace with your parents. Shift your relationship with your parents to a healthy adult-adult relationship.

2). Find people you want on your crew. Ask them to train with you and help hold each other accountable for your greatness.

3). PAY IT FORWARD Leap! Launch that new project, open your company, start your podcast, write your book or your screenplay, build your rocket, take the next step towards the future of your dreams. Then invite people who are inspired by you to start their own Hero's Journey today.

Who Else Do You Want on Your Crew?

If you are willing to hold yourself to a higher standard and keep doing the work of embodying the "New Right Stuff," write your name below and pass this book on to someone else you respect and want on your crew with you. Or write the names of the people you want on your crew and get them their own copies to mark up, take notes in and keep handy!

____

____

____

____

____

Thanks for being part of the journey.

Follow Loretta on Twitter (@lorettawhtsides) or connect via Instagram (@lorettahidalgowhitesides), Facebook (Loretta Hidalgo Whitesides), YouTube (Loretta Whitesides) and LinkedIn (Loretta Hidalgo Whitesides).

You can also check out the website www.TheNewRightStuff.org

Acknowledgements

This book would not have been possible without the unconditional support of my extraordinary husband George T. Whitesides. Thank you for all the magic you bring to life (and for your edits).

Special thanks also to my mother-in-law Barbara Whitesides for copy editing the entire book and being such an indefatigable support to our household. And to Ron Rosano, Bill Wesler, Melissa Sampson and others who submitted their comments and feedback on early drafts. Thank you to my parents for giving me this great life (and watching my kids during many trips!) and my children for putting up with me being "busy" for too often.

To Karen Lau, Kelly Snook, Jessy Schingler and the 4D crew—thank you for the inspiration to follow my voice and believe I have something worth expressing. Thank you for the practice and the listening (and for pushing me to be my best self.)

I would also like to thank my many extraordinary school teachers from Sr. Ursula, to Mrs. Ryan, Mr. Rutherford, Fr. Carolyn, and Mrs. Allen. I am forever indebted by your generosity. To my amazing mentors, Dr. Chris McKay, Michael Eastwood, Rick Tumlinson, Mark Stave, Evan Unruh, Frank White, Alan Cahn, Dr. Peter Diamandis and most of all Richard Condon. I am very grateful for the transformational courses and coaching I have benefitted from so greatly. Because of you I have the life of my dreams and get to make the difference I came to Earth to make.

Thank you to my Camp St. Michael and Redwood Adventure Camp family. That is where I first learned how to be myself and how to share myself. I am forever grateful especially to Dina McDonald, Bob Bailey and Mary Kirby.

Thank you to my NASA Johnson Space Center crew for all the adventures and support. Thank you to all the Yuri's Night community for all that you do to share our vision of using space to bring the world together especially Mike Mongo, Chris Lewicki, Christy Fair and Tim Bailey who are always there for me. Thank you to all the Jedi at Virgin Galactic and Virgin Orbit and all the SEDS, ISU, and SpaceGen people from around the planet for your support.

And a very special thank you to Kyle Schember of Subtractive for recording and editing all the audio for the Audible version of this book! You are an awesome partner and a visionary space producer. Thank you to my dear Leap friend Jon Marro for creating all the beautiful illustrations in this book, Marcel Kaufmann for coining the term SpaceKind and graciously allowing us to develop it and to Rachel Lyons and Quade MacDonald for help with the manuscript.

I owe Julia Cameron for her inspiration in *The Artist Way* (that I first read over 20 years ago), for the dimensions of this workbook and the wide margins to allow for lots of note taking. I owe Jon Marro's *Keepers of Color* for the opening invitation to the journey!

Thank you to all the muses, the Universe, and God, for continuing to send love and inspiration. I hope you will be still and quiet so you can hear their gifts for you too.

Resources

Recommended Reading

Fiction

Illusions: The Adventures of a Reluctant Messiah, By Richard Bach
Hitchhikers Guide to the Galaxy, By Douglas Adams
Parable of the Sower, By Octavia Butler
Stranger in a Strange Land, By Robert Heinlein

Nonfiction

The Overview Effect, By Frank White
The Surrender Experiment, By Michael Singer
How to Win Friends and Influence People, By Dale Carnegie
The Artist's Way, By Julia Cameron
The Way of Integrity, By Martha Beck

Courses

The Curriculum for Living at Landmark Worldwide
Date With Destiny and Unleash the Power Within by Tony Robbins

Digital Tools

Five Minute Journal App

Index

Abbey, George, 159

accepting compliments, 110

acknowledgement, 111, 113

aggression, 71

Aldrin, Buzz, 82, 185

analysis paralysis, 66

anger, 17, 18, 37, 67, 69, 75, 96, 104, 125

Apollo, 25

apologizing, 28, 34, 80, 91, 132

arrogance, 71

Ask for Help, 9

Bach, Richard, 45, 184

Be like a cat, 27

being on time, 104

Bleiler, Gretchen, 75

blind spots, 14, 15

Blue Origin, 57, 58, 132

Bohm, David, 148

Bohm Dialogue, 148

Branson, Richard, 81

Butler, Octavia, 46

calling, 6

Campbell, Joseph, 19, 163

Carnegie, Dale, 45, 60, 73

challenging yourself, 29

coaches, 46

Coehlo, Paulo, 45

Colbert, Stephen, 146

Connector, 141

Cox, Kenneth J., 158

criticism, 73, 74

daily action, 47

Dalai Lama, 152

dark side, 52, 53, 54

Desiderata, 77

dissatisfaction, 63

Dream Big, 7, 167

failure, 10, 62, 70, 72, 73, 122, 147, 154, 156

fear, 53, 69

Frankl, Viktor 45

Frost, Robert, 86

furballs, 77, 79

Gagarin, Yuri, 173

Gladwell, Malcolm, 141

Gupta, Anurag, 64, 73

Hero's Journey, 19, 48, 117, 163

honeymoon period, 64

Hoover, Bob, 122

humor, 97

Imposter Syndrome, 52

inclusive language, 99

integrity, 3, 68, 92-96, 99, 115, 151

It's Not About the Nail, 31

jealousy, 75

Jedi, 3, 13, 17-19, 28-30, 32, 35, 37, 38, 41,-43, 56, 75, 77, 90, 96, 104-107, 110, 114, 115, 128, 136, 139, 142, 143, 147, 152, 159, 182, 184

Jemison, Mae, 27

Jiang, Jia, 69

Lincoln, Abraham, 73

listening, 30, 35

Logan, Dave, 69

Lucid, Shannon, 152

Magnificent Desolation, 82

Mailer, Norman, 181

Mars, 11, 51

Masten, 58

McKay, Chris, 142, 144

Musk, Elon, 57, 58

NASA, 3, 7, 27, 57, 69, 72, 82-85, 100, 106, 142, 144, 148, 153- 158, 176, 177, 181, 190

Never Give Up, 10

New Right Stuff, 4, 24, 189

Pan-Industry Jedi Council, 114

patience, 19, 37

Penn, William, 68

Plato, 68

Pressfield, Steven 45

Privilege is Blind, 5, 100

remembering people’s names, 106, 108

Ride, Sally, 85

Ringer, Judy, 128

Roddenberry, Gene, 26, 140, 157

Rohn, Jim, 14

Ross, Tamaira, 132

Ruiz, Don Miguel, 89

Sarcasm, 61

Scott, Winston, 107

Sierra Nevada, 58

SpaceX, 57, 58, 85

Speaking, 32, 34, 36

Star Trek, 26, 39, 156

Star Wars, 3, 17, 39, 42, 114

STEAM, 157

Stress, 62

suicide, 82

taboos, 81, 82, 85

Taming of the Shrew, 34

The Empire Strikes Back, 53

The Force, 42, 51

The Four Agreements, 89, 90

tone, 33

Trungpa, Chogyam, 57

Virgin Galactic, 58, 85, 97, 190

Virgin Orbit, 75

von Braun, Wernher, 25

Whitesides, George, 34, 97, 190

Wooden, John, 60

Worden, General Pete, 73

Yoda, 17, 19, 37, 38, 42, 53, 151, 184

Yuri's Night, 103, 173

Zander, Benjamin TED Talk, 33

Loretta Hidalgo Whitesides and her husband George are 'Founder Astronauts' with tickets for a sub-orbital spaceflight on Virgin Galactic's Spaceship 'Unity.' Trained as an astrobiologist at Stanford and Caltech, Loretta has been to the Arctic to study plant life in extreme environments and to the hydrothermal vents at the bottom of the ocean with *Avatar* director James Cameron to film the IMAX, *Aliens of the Deep*. She has worked at multiple NASA centers, as well as the X PRIZE Foundation and Wired.com.

She is the Co-Creator of Yuri's Night, The World Space Party, celebrated annually around April 12. Loretta has over five hours spent floating weightless (30 seconds at a time) on a 727 aircraft as a Flight Director for Zero-G. Loretta's passion is helping people fulfill what they came to Earth to do and helping humanity become a species we would be proud to send to the nearest star system.

Made in the USA
Columbia, SC
23 November 2021